Piping and Pipe Support Systems

Piping and Pipe Support Systems

Design and Engineering

Paul R. Smith, P.E.

Thomas J. Van Laan, P.E.

Boston, Massachusetts   Burr Ridge, Illinois
Debuque, Iowa   Madison, Wisconsin   New York, New York
San Francisco, California   St. Louis, Missouri

Library of Congress Cataloging-in-Publication Data

Smith, Paul R.
  Piping and pipe support systems.

  Bibliography: p.
  Includes index.
  1. Pipe lines—Design and construction. 2. Pipe
supports—Design and construction. I. Van Laan, Thomas J.,
II. Title.
TJ930.S625 1987   621.8′672   86-18628
ISBN 0-07-058931-3

McGraw-Hill

A Division of The McGraw·Hill Companies

Copyright © 1987 by McGraw-Hill, Inc. All rights reserved.
Printed in the United States of America. Except as permitted
under the United States Copyright Act of 1976, no part of
this publication may be reproduced or distributed in any
form or by any means, or stored in a database or retrieval
system, without the prior written permission of the publisher.

8 9 10 11 12 13 14 BKMBKM 9 9 8 7

ISBN 0-07-058931-3

This book is printed on acid-free paper.

*The editors for this book were Betty Sun and Janet B. Davis, the
designer was Naomi Auerbach, and the production supervisor was Annette
Mayeski. It was set in Century Schoolbook by University Graphics, Inc.*

This book is dedicated to those who gave us the support necessary to complete our work: Ruth Dreessen, Belinda Hunter-Taylor, Wade Larson, and Genevieve Smith.

Contents

Preface   xi
Acknowledgments   xv

Chapter 1. Piping Systems and Power Plant Evolution   1

1.1	Introduction	1
1.2	Historical Overview of Piping System Engineering	5
1.3	Overview of Thermoelectric Power Plants	7
1.4	Commercial Nuclear Reactor: Overview	12

Chapter 2. Codes, Standards, and Regulations   17

2.1	Introduction		17
2.2	History of Applicable Documents (Codes, Standards, Recommended Practices, and Guides)		18
2.3	Overview of Applicable Documents Relating to Piping		21
	2.3.1	American Institute of Steel Construction	21
	2.3.2	American National Standards Institute	22
	2.3.3	Manufacturers Standardization Society of the Valve and Fittings Industry	23
	2.3.4	American Society of Mechanical Engineers	23
	2.3.5	Jurisdictional Boundaries	27
	2.3.6	American Society for Testing and Materials	29
	2.3.7	Uniform Boiler and Pressure Vessel Laws Society	29
	2.3.8	National Board of Boiler and Pressure Vessel Inspectors	30
	2.3.9	American Society of Heating, Refrigeration, and Air Conditioning Engineers	30
	2.3.10	Pipe Fabrication Institute	30
	2.3.11	Occupational Safety and Health Administration	30
	2.3.12	American Welding Society	31
	2.3.13	Building Codes	31
	2.3.14	Nuclear Regulatory Commission	31
	2.3.15	Technical Publications	35
	2.3.16	International Standards	36
	2.3.17	International Atomic Energy Agency	36
	2.3.18	Atomic Industrial Forum	37
	2.3.19	Institute of Nuclear Power Operations	37
	2.3.20	National Fire Protection Association	37
	2.3.21	American Nuclear Society	37
2.4	Role of the Quality Assurance Organization Within the Nuclear Industry		38

viii   Contents

Chapter 3. Technical Piping Documentation — 40

- 3.1 Introduction — 40
- 3.2 Flow Diagrams, Line List, and Design Specifications — 42
- 3.3 Piping Layout — 46
- 3.4 Piping Design Drawings — 49
- 3.5 Composite Drawings and Scale Models — 52
- 3.6 Piping Isometric Drawings — 53
- 3.7 Stress Isometrics — 55
- 3.8 In-Service Inspection Drawings — 58

Chapter 4. Overview of Pipe Stress Requirements — 61

- 4.1 Introduction — 61
- 4.2 Theoretical Basis for Piping Codes — 62
  - 4.2.1 Failure Theories — 62
  - 4.2.2 Stress Categories — 63
  - 4.2.3 Stress Limits — 64
  - 4.2.4 Fatigue — 66
- 4.3 Occasional Loads and Service Levels — 68
- 4.4 B31 Committee Projects — 69
  - 4.4.1 B31.1 Power Piping Code — 70
  - 4.4.2 B31.3 Chemical Plant and Petroleum Refinery Piping Code — 72
  - 4.4.3 B31.7 Nuclear Power Piping Code — 76
  - 4.4.4 B31.8 Gas Transmission and Distribution Piping Code — 77
  - 4.4.5 ASME Boiler and Pressure Vessel Code, Section III, Subsection NB — 78
  - 4.4.6 ASME Boiler and Pressure Vessel Code, Section III, Subsections NC and ND — 82

Chapter 5. Piping Design Loads — 85

- 5.1 Introduction — 85
- 5.2 Sustained Loads — 87
  - 5.2.1 Sustained Loads—Weight — 87
  - 5.2.2 Sustained Loads—Pressure — 98
- 5.3 Occasional Loads — 102
  - 5.3.1 Occasional Loads—Wind — 103
  - 5.3.2 Occasional Loads—Relief Valve Discharge — 109
  - 5.3.3 Occasional Loads—Seismic — 112
  - 5.3.4 Vibration — 127
- 5.4 Expansion Loads — 128
  - 5.4.1 Determination of Thermal Loads and Stresses — 128
  - 5.4.2 Determination of Thermal Movements — 134
- 5.5 Load Combination — 150
- 5.6 Computer Calculations — 150

Chapter 6. Pipe Support Hardware — 151

- 6.1 Introduction — 151
- 6.2 Weight Supports — 153
  - 6.2.1 Rigid Supports — 153
  - 6.2.2 Variable-Spring Supports — 161
  - 6.2.3 Constant-Spring Supports — 164

6.3	Rigid Restraints	169
6.4	Snubbers	182
6.5	Sway Braces	186
6.6	Baseplates	187
6.7	Multiple-Pipe Restraint Frames	189

Chapter 7. Piping Support Design Process — 191

7.1	Introduction	191
7.2	Preparation of Information and Data Required for Pipe Support Design	193
7.3	Selection of Pipe Support Locations	195
7.4	Marking Support on Isometrics	199
7.5	Pipe Stress Analysis	200
7.6	Determination of Support Design Loads	202
7.7	Pipe Support: Conceptual Design	203
7.8	Pipe Support: Detail Design	205
7.9	Detail Design Drawing and Bill of Materials	208
7.10	Checking	208
7.11	Fabrication, Installation, and Verification of As-Built Configuration	210

Chapter 8. Manual Calculation Methods — 211

8.1	Introduction		211
8.2	Strength of Materials		212
	8.2.1	Stress-Strain Relationship	212
	8.2.2	Axial Stresses	213
	8.2.3	Bending Stresses	218
	8.2.4	Shear Stress	220
	8.2.5	Deflection	222
8.3	Support Analysis		225
	8.3.1	Local Stress Evaluation	225
	8.3.2	Evaluation of Vendors' Hardware	232
	8.3.3	Structural Analysis of Support Steel	232
	8.3.4	Evaluation of Structural Attachments	240
8.4	Sample Problems		253

Chapter 9. Computer Applications for Design and Analysis — 273

9.1	Introduction		273
	9.1.1	Introduction to the Computer	274
	9.1.2	Input/Output Devices	277
	9.1.3	Central Processing Unit	280
9.2	Computer-Aided Design and Drafting		281
9.3	Computer-Aided Engineering		285
	9.3.1	Introduction	285
	9.3.2	Input Methods for Computer-Aided Engineering	291
	9.3.3	Sample Structural Analysis Problem	315
	9.3.4	Sample Pipe Stress Problem	326

Index   327

Preface

This text covers the design and analysis of piping systems. Although the topic has been approached with special attention to the support of these systems (primarily owing to a perceived lack of material on the subject), the full scope of this book includes power, industrial, and chemical applications, and goes beyond the piping system to show the role of piping design criteria in the setting of the plant as a whole. The intention has been to explore the theoretical basis behind, and the regulatory requirements governing, piping and support design and analysis, as well as to present a practical view of the subject for the engineer working in the field.

Although the fields of piping and support engineering have a long tradition, they have undergone significant change within the last few decades. Increasing desires to optimize safety and economic considerations, as well as new forms of technology, have led to major changes in piping engineering. To protect the public (especially in the nuclear and chemical industries), numerous international, federal, state, and local regulations have been enacted since the late 1960s. In response, various professional groups have established codes and standards governing piping design. The increasing concern for safety has forced the development of new design criteria aimed at ensuring the integrity of piping and its supports under unusual loading conditions, such as those imposed by earthquakes, pipe ruptures, or in-line equipment loads. The traditional approaches to piping design were inadequate to meet these new demands; thus new analytical tools were required and have been developed during the last few years, usually as an extension of the growing power of computers. In such cases, the art of piping design has become a task for the computer-aided engineer.

The authors recognize both the value of the traditional methods and the inevitability of engineering progress. For normal operating conditions, design loads are usually low in magnitude and may require only simple hardware for support. Circumstances such as these may usually be adequately addressed by traditional design methods, which the text therefore systematically and completely presents. In addition, this book discusses the application to piping design of such state of the art methods as computer-aided design, drafting, and engineering.

The discussion herein of codes and standards gives this book a value beyond that of a simple piping design handbook. The text emphasizes the documents associated with the power industry, especially those associated with the nuclear power industry, because they provide the most comprehensive set of technical and documentation requirements. These requirements in most cases are more severe than those of fossil fuel, hydroelectric, industrial, and commercial piping applications, and can be said to "envelope" them. It is important to note that the evaluation of applicable documents (i.e., codes, standards, guides, regulations, recommended practices, etc.) is an ongoing process. Thus, this book addresses those documents that are viewed as significant to the subject at the time of publication. Also, only the latest edition of a document is discussed here, except in cases where an earlier version has historical or other significance. Since design standards provide the minimum guidelines, and often cover a broad range of applications, some have been modified or have had portions omitted by the authors in order to emphasize those areas considered significant to piping and support design. Because of this, the reader is urged not to take the portions of the codes quoted herein out of context, but rather to refer to the actual code publications.

The principles of piping and support design and analysis are explained using both visual aids and sample calculations, which will serve to clarify specific criteria as well as problems normally encountered by the piping practitioner. The text is also designed to serve as a reference handbook for engineers and designers currently working in the piping industry, at power, process, or industrial facilities. Since the material covered provides a central source for data and methodology, it is hoped that engineers and designers will use this text as a daily reference and design tool. Therefore, a collection of engineering design data is extracted from commonly used sources and reprinted herein. Additionally, the design and analysis procedures are explained, along with background information to aid the design engineer in altering the method for application to other situations one may encounter. Methods presented here not only emphasize analysis techniques, but equally stress the various types of commercially available hardware used in the industry, along with their selection criteria.

The material is presented in what the authors believe to be the most logical order of study for a reader with minimum foreknowledge of piping and support engineering. Readers may read only those sections that best suit their needs.

The first three chapters serve as an introduction to the engineer unfamiliar with the field. Chapter 1 introduces the subject of piping systems and engineering applications; Chap. 2 comprises historical and technical coverage of the various codes and regulatory bodies influencing piping design; and Chap. 3 describes the different design drawings and other technical documentation required when designing piping systems. The next two chapters deal with piping loadings: Chap. 4 outlines pipe stress requirements, and Chap. 5 covers the determination of support loads.

The final four chapters cover the actual details of piping and support design. Chapter 6 describes the various types and uses of pipe support hardware; Chap. 7 details the steps of the support design process, from initial support location through the design, fabrication, and installation stages; Chap. 8 covers manual calculation techniques for the analysis of the pipe pressure boundary, hardware capacity, structural steel, welding, and baseplates; and Chap. 9 explains the use of the computer in engineering design, drafting, and analysis. Sample computer calculations for pipe stress and structural supports are presented.

Although the material has been arranged to be of maximum use to the reader, it must be stressed that equations and data should not be used for safety-related design without first securing competent advice from a professional engineer with respect to the suitability of the particular application. While the material is believed to be technically correct, the authors do not warrant it suitable for any general use. The same recommendation that was made earlier with respect to codes continues to apply: the reader is urged to refer directly to the actual sources quoted for any restrictions placed upon the use of the techniques.

This book originated from a set of student lecture notes accompanying a series of design and analysis seminars which the authors conducted over 100 times, at electric utilities, for architects and engineers, and at technical training centers in six major cities throughout the United States. The material incorporates feedback from the students attending those seminars. It is hoped that educational groups such as colleges, universities, and training centers will consider using this book as the basis for a technical course on piping and support design and engineering. The subjects covered—statics, strength of materials, structural design, and computerized design and engineering—will give the student practical use of theory studied in other portions of the academic curriculum.

Much as this book has benefited from the feedback from those attending previous seminars, the authors realize it could benefit still further from readers' comments. The authors therefore welcome suggestions for improvements for future editions.

Paul R. Smith
Thomas J. Van Laan

Acknowledgments

The authors would like to thank the following companies and organizations for the assistance they provided during the preparation of this text:

American National Standards Institute
American Society of Mechanical Engineers
American Welding Society
Anchor/Darling Industries, Inc.
Bergen-Paterson Pipesupport Corp.
Construction Systems Associates, Inc.
Control Data Corporation
Corner & Lada Co., Inc.
Engineering Planning and Management, Inc.
General Electric
IBM Corp.
Impell Corporation
Intergraph Corp.
ITT Grinnell Corp.
James F. Lincoln Arc Welding Foundation
Manufacturers Standardization Society of the Valve and Fittings Industry
McDonnell Douglas Information Systems Co.
Northeast Utilities
Nuclear Power Services, Inc.
Owens-Corning Fiberglas Corp.
Power Piping Co.
PRS Industries, Inc.
Tetracom Services
Welding Research Council
Westinghouse Electric Corporation

Certain pages in this book have been reproduced from copyrighted ITT Grinnell Corporation Literature with the permission of ITT Grinnell Corporation.

Chapter

1

Piping Systems and Power Plant Evolution

1.1 Introduction

A *pipe* can be defined as a tube made of metal, plastic, wood, concrete, or fiberglass. Pipes are used to carry liquids, gases, slurries, or fine particles. Pipes combined into networks have been in use since prehistoric times, originally for irrigation purposes.

A piping system is generally considered to include the complete interconnection of pipes, including in-line components such as pipe fittings and flanges. Pumps, heat exchangers, valves, and tanks are also considered part of the piping system. The contributions of piping systems are essential in an industrialized society—they provide drinking water to cities, irrigation water to farms, and cooling water to buildings and machinery. Piping systems are the arteries of our industrial processes; they transmit the steam to turn the turbines which drive generators, thus providing the electricity that illuminates the world and powers machines.

Figure 1.1 illustrates the magnitude of piping required in a typical chemical process plant. Piping systems account for a significant portion of the total plant cost, at times as much as one-third of the total investment.

Figure 1.2 is an artist's illustration of another type of plant, a power plant that burns fossil fuels. Figure 1.3 illustrates a computerized model of a nuclear power plant, showing the piping and major equipment. These types of plants require an enormous amount of piping within a smaller area than the chemical process plant of Fig. 1.1. Piping systems arranged within a very confined area can be an added challenge to piping and support engineers.

The initial design of a piping system is established by the functional requirements of piping a fluid from one point to another. The detail

Figure 1.1 Partial view of piping system in representative chemical process or refining plant.

design is affected by such criteria as the type of fluid being transported, allowable pressure drop or energy loss cost (pumping power), desired velocity, available materials of construction, etc.

In the design of piping systems and their supports, the factors that need to be taken into consideration depend on the type of plant in which the piping will be installed. As noted, piping systems constitute a major expense in process and power plants, so economy is usually the major consideration. An exception occurs during the design of piping and restraints for nuclear power plants, when the minimization of environmental hazard becomes the premier motivating factor. For nuclear submarine piping systems, space limitations may be of critical importance.

The supporting of piping systems requires a significant engineering, design, fabrication, and installation effort. A typical 600-MW fossil fuel power plant has approximately 4000 pipe supports, while a typical 1100-MW nuclear power plant has over 10,000 pipe supports and restraints. In some cases, special superstructures must be built solely for the purpose of supporting piping systems.

This book will review piping system design evolution in thermoelectric power plants as well as all relevant codes, standards, design data, and criteria used to design, engineer, and fabricate piping systems and their supports.

This book both describes the engineering bases and regulatory requirements influencing the development of pipe stress analysis and pipe support techniques, and shows their role in the design of piping systems and

Figure 1.2 Representative outdoor fossil fuel power plant.

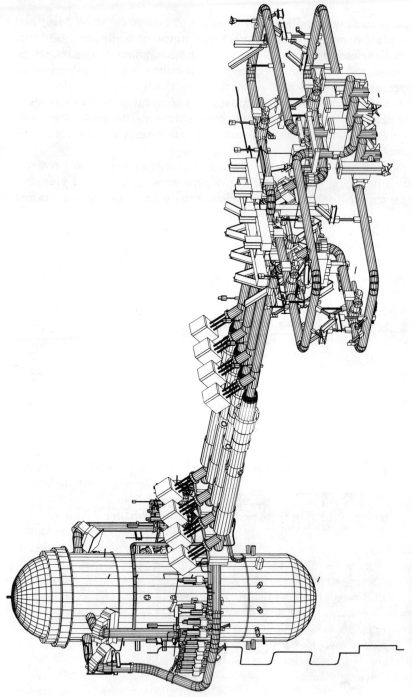

Figure 1.3 Computerized model of nuclear power plant piping, vessels, and equipment. (*Courtesy of Construction Systems Associates, Inc., Marietta, Ga.*)

the plant as a whole. The discussion of codes and standards makes this text useful for instructional and reference purposes.

The text describes the technical basis for and practical approaches to the design and analysis of piping systems with special attention to supporting these systems within defined, allowable criteria. The principles of piping design and analysis are explained through visual aids (graphs, charts, and illustrations) and sample calculations (methods and examples), by using manual estimates and computer simulation of the piping system.

1.2 Historical Overview of Piping System Engineering

When wind, water, and muscle were the prime movers of western civilization, the basic tools used by humans were windmills, wagons, and pails. The advent of the industrial revolution, especially the practical use of steam power in the 1700s, required the design and manufacture of piping to withstand the rigors of conveying pressurized and heating fluids. The combination of very high pressures, thermal stresses, and thermal deformations required that fundamental design requirements and analytical techniques be developed. However, piping system design progressed with little or no design standards or code limitations during the early years of the industrial revolution.

The electrical revolution triggered a spiraling acceleration in U.S. energy use. Even though electricity is only one form of energy—and despite the fact that it cannot be easily stored—its convenient use made it the preferred form of the booming U.S. industry. The introduction, to meet the electrical demand, of turbine plants (1907 to the 1920s) with superheated steam [temperatures reaching 600°F (316°C) and gauge pressures of 300 psi (2 MN/m^2)] posed the next major piping system design challenge. These design conditions exceeded safe cast-iron design values [450°F (232°C)], thus requiring the introduction of cast steel for critical components. As late as 1923, the wisdom of higher pressure was still being debated, but the trend toward higher design requirements had already begun. By 1924, the steam gauge pressure had increased to 600 psi (4 MN/m^2), doubling in just a few years. One year later, steam pressures and temperatures of 1200 psi (8 MN/m^2) and 700°F (371°C) were achieved, demonstrating the advances made in the development of the steam generator and attached piping.

By 1957, some 1100°F (593°C) designs were in service with 1200°F (649°C) designs projected, using austenitic stainless-steel materials in the high-temperature zones. Currently, the top gauge pressure is 2400 psi (16 MN/m^2) for most fossil fuel plants. With new materials available, the boiler, turbine, and piping have equal strength capabilities.

The increase in operating temperatures and pressures led to the development of the ASA (now ANSI) B31 Code for Pressure Piping. During the 1950s, the code was segmented to meet the individual requirements of the various developing piping industries, with codes being published for the power, petrochemical, and gas transmission industries among others. However, the greatest impetus to the standardization, codification, and resulting documentation of piping system design occurred in support of the U.S. nuclear power industry to assure the maintenance of public safety. The 1960s and 1970s encompassed a period of development of standard concepts, requirements, and methodologies. The development and use of computerized mathematical models of piping systems (such as shown in Fig. 1.4), have brought analysis, design, and drafting to new levels of sophistication. The ability to predict and simulate the piping system's response under numerous potential loading conditions brought the design of piping systems into the twentieth century. The ASME nuclear piping code has introduced higher levels of understanding of strength of materials theory to piping design requirements. (For more information on the history of piping, see the history of the ASME code in Chap. 2.)

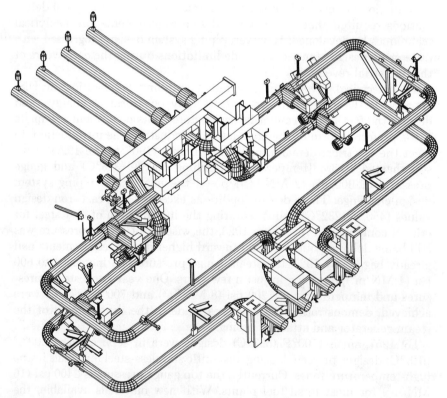

Figure 1.4 Computerized model of piping systems. *(Courtesy of Construction Systems Associates, Inc., Marietta, Ga.)*

To protect the public, numerous international, federal, state, and local regulations have been enacted over the past 20 years. In response, various professional groups have established codes and standards governing piping design. This text collects and organizes this data in such a way that a piping engineer can refer to those codes and standards that apply to a specific industry or application.

The text focuses on both traditional and recently developed engineering activities required to ensure the integrity of piping design. Pipe stress analysis, for example, ensures the safe operation of piping systems by verifying their structural and pressure-retaining integrity under design loading conditions postulated to occur over the lifetime of the installation.

Traditionally, this has been accomplished by making a mathematical model of the piping system which is then subjected to loading conditions defined in the design specification. For normal operating conditions, design loads are usually low in magnitude and require only simple hardware to support. The traditional stress analysis method is adequate to address design loads of this nature. Because these remain useful, the book systematically and completely presents traditional methodologies for performing stress analysis.

However, increasing concern for safety has forced the development of design criteria for ensuring the integrity of piping and pipe supports under unusual loading conditions, such as those imposed by earthquakes, pipe ruptures, or in-line equipment loads. The greater complexity and magnitude of these design criteria require tools for stress analysis that have been developed only in the past 10 years. The art of piping design has become the bailiwick of the computer-aided design engineer. The use of computers allows more accurate determination of loading distribution throughout the piping and supporting structures, making computer-aided design the preferred methodology for most applications. Because of the increasingly important role of computers in piping design and analysis, the book emphasizes these newer methodologies for performing stress analysis.

1.3 Overview of Thermoelectric Power Plants

Most of the electric power in the United States is currently produced in steam plants using fossil fuels and high-speed turbines. The particular type of steam plant selected is usually based on economic considerations stemming from three principal elements:

1. Capital equipment cost
2. Fuel cost
3. Operating and maintenance cost

The capital equipment cost includes the cost of such major components as the steam generator, turbine, electric generator, condenser, feedwater heaters, pumps, fuel and waste handling facilities, piping, structures, and real estate. The fuel cost is calculated based on the average anticipated cost during the plant lifetime. Such future costs are often difficult to estimate, since social and political events, such as the oil embargo of the early 1970s or the decontrol of domestic oil prices, may suddenly and radically affect them. The capitalized cost of the plant must be balanced against the ongoing cost of fuel operation and maintenance as well as the future cost of decommissioning the plant. For example, a nuclear power plant is distinguished by its significantly higher capital cost and significantly lower fuel cost than would be found in a fossil fuel plant.

Figure 1.5 illustrates how electricity is generated in a typical fossil fuel thermoelectric generation station. The major components, consisting mainly of the capital equipment previously listed, are as follows:

1. *Boiler.* A container into which water is fed and heat is applied by burning fuel is a boiler. This heat changes the water into steam much as a teakettle on a stove.

2. *Turbine and generator.* The *turbine* is a device used to change the energy in steam to motion, much as a windmill uses air. A turbine consists of blades mounted on a shaft that turns. In most fossil fuel stations, this shaft rotates at 3600 rpm. The *generator* is a device which produces electricity. It has two parts, a stationary part made of coils of copper wire and a rotating part that is a magnet connected to the turbine shaft.

3. *Condenser.* A container where the steam used in the turbine is changed back to water is a *condenser*. It accomplishes this by circulating cool water inside tubes which causes the steam to cool and condense on the outside of the tubes, much as misty water forms on a mirror after a hot shower. This water is called *condensate*. The condenser is a set of tubes kept cool so that the steam condenses into water. The condenser tubes are cooled by a separate circuit of cooling water, as shown in Fig. 1.5.

4. *Hotwell and condensate pump.* The *hotwell* is a reservoir at the bottom of the condenser which provides a place for the condensate to collect. The *condensate pump* is a mechanism that moves the condensate from the hotwell toward the boiler.

5. *Feedwater heaters.* For more efficiency, these devices heat the condensate before it enters the boiler so that less fuel is used in the boiler. To do this, the heaters use some of the same steam that turned the turbine. This steam is removed before it enters the condenser and thereby reduces the amount of heat lost to the condenser cooling water.

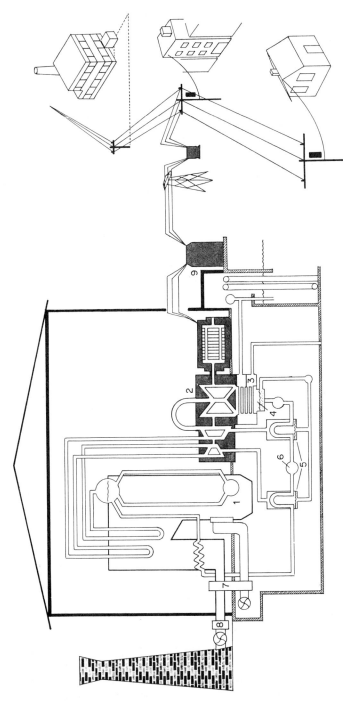

1 Boiler   4 Hotwell   7 Air Preheater
2 Turbine   5 Feedwater Heaters   8 Precipitator
3 Condenser   6 Boiler Feed Pump   9 Screen House

Figure 1.5 Typical thermal electric power plant. (*Courtesy of Northeast Utilities.*)

6. *Boiler feed pump.* The main pump that returns the condensate and feedwater to the boiler at very high pressure is the boiler feed pump.
7. *Forced-draft fan and air preheater.* The *forced-draft fan* is a large fan that pushes air into the furnace of the boiler to help burn the fuel. The *air preheater* helps the fuel to burn better by warming the air before it enters the boiler.
8. *Precipitator, induced-draft fan, and stack.* The *precipitator* is a device used to remove and collect particles left over from the burning of a fuel after leaving the furnace. The *induced-draft fan* is a large fan that pulls burned gases out of the boiler. The *stack* is simply a tall chimney.
9. *Screenhouse.* The *screenhouse* is a building located near a source of cooling water (i.e., river or ocean) which houses the following: the *trash rack*, which prevents large items, such as logs, from entering the building; the *revolving screens*, which remove smaller items such as twigs, seaweed, and fish from the cooling water; and the *circulating pumps* which force the cooling water through the condenser.

Simply expressed, the generation of electricity begins when water is heated by the burning of fuel to form steam. The steam passes through the turbine, causing the turning of the rotor axle connected to it. When the rotor turns, it spins a coil of wire within a magnetic field, producing electricity. This electricity must be delivered through transmission wires to the user as soon as it is generated since electricity cannot be stored in a commercially economical manner.

The fuel used in Fig. 1.5 is not identified: power plants may use coal, natural gas, nuclear, or even refuse as fuels. The steam generators shown in Fig. 1.5 could also use the heat of fission (splitting of atomic particles) from nuclear fuel to boil the water. A nuclear power plant, therefore, produces electricity in much the same way as a fossil fuel plant, except that the fuel is fissionable material (e.g., uranium or plutonium). Note that there is only nuclear heat; there is no such thing as nuclear electric power. Heat, once generated, regardless of fuel type, is used to generate electricity in the same way whether in a fossil fuel or nuclear power plant.

For a fair comparison of energy content, the cost of energy available from fissionable fuel must be increased, compared to fossil fuel, by a factor to account for the additional engineering, design, licensing, and decommissioning efforts associated with nuclear power. Since nuclear power plant requirements for engineering and design (including piping) are far more stringent than those associated with fossil fuel power plants, the overall cost of nuclear power may exceed the cost of fossil fuel power.

Fossil fuels might be abundant today; however, some face depletion within the foreseeable future. Alternative energy sources such as nuclear power and hydropower are needed while more attractive energy sources (solar, fusion, etc.) are being developed.

Most of the electric power currently used in the United States is produced by generating plants using fossil fuels and high-speed turbines. They utilize steam-driven generators of 60-MW to 1300-MW capacity supplied with steam from boilers generating 500,000 to approximately 10,000,000 lb/h (1,100,000 to 22,000,000 kg/h).

The 1973–1974 oil embargo threw the world's energy balance into complete disarray. It put numerous strains on both the economic and technical resources of industrialized nations. The embargo seemed to reverse, at least temporarily, the ever-increasing need for electricity in the industrialized nations.

Like oil, natural gas is also prey to unpredictable price explosions, and like oil, it is a depletable, finite energy source. Although recent estimates indicate that our natural gas reserves may be larger than previously thought, the cost of recovery is uncertain and may be prohibitive. Finally, oil and natural gas are vital feedstores in the production of plastics, other petroleum products, fertilizer, and pharmaceuticals. These other needs for oil and natural gas require consideration in evaluating the various sources of energy for the future.

We cannot eliminate uncertainty from our attempts to see into the future. Still, as the U.S. population and economy grow, there is good reason to believe that the demand for electric energy will grow too. Industry, the marketplace, transportation, agriculture, and life at home will be heavily influenced by the reliability and sufficiency of our electrical supply. To help maintain that supply, conservation, electricity produced by coal, and nuclear electricity are reported to be necessary.

The 1979 Three Mile Island nuclear power plant accident caused the United States to stop and reevaluate its attitude toward its nuclear power industry. However, energy generation professionals still believe that nuclear power is indispensable to industrialized nations.

A survey of U.S. scientists was conducted in 1980, shortly after the Three Mile Island accident. The detailed questionnaire on energy issues was sent to a random sample of (1) scientists listed in *American Men and Women of Science,* (2) scientists in energy-related disciplines, and (3) scientists in fields closely related to nuclear energy. The scientists were asked to name the energy sources which were expected to make the greatest contributions to our needs by the year 2000. The questionnaire listed 16 possible sources, ranging alphabetically from biomass to wind power. Table 1.1 summarizes the responses to the questionnaire, relative to six energy sources. Most respondents viewed coal as the primary energy source of the future, followed by oil, natural gas, nuclear fission, and conservation.

TABLE 1.1 Tabulated Responses to Question: "What Resources Will Make Major Contributions to Our Energy Needs?"

Energy source	Antinuclear groups, %	Industry, %	Financiers, %	NRC, %	Other regulators, %	Congress, %	Outside experts, %
Coal	58	96	78	94	75	91	95
Oil	50	57	50	63	84	67	79
Natural gas	42	44	41	34	48	52	67
Nuclear fission	0	52	24	28	52	25	33
Solar heat	42	1	0	3	12	10	2
Conservation	100	16	29	19	50	52	38

SOURCE: Data from *Nuclear Electricity—Who Stands Where. (Courtesy of Dr. S. Robert Lichter and Dr. Stanley Rothman.)*

Nevertheless, the utilization of nuclear power continues to grow worldwide. This demonstrates that nuclear power will continue to be a primary contributor on the global scene, regardless of its immediate future in the United States.

1.4 Commercial Nuclear Reactor: Overview

Many different nuclear power reactor designs are commercially available. It has been reported that at least 1200 combinations of reactor type variables exist. The reader is referred to other texts for some of these variables and combinations.

Nuclear power for peaceful use began in the United States in 1957 with the successful operation of a 60-MW unit at Shippingport, Pennsylvania. By the mid-1960s, two types of commercial reactors—the boiling-water reactor (BWR) and the pressurized-water reactor (PWR)—had evolved as the two main types of reactors in the United States. Approximately 80 percent of the world's 521 nuclear plants reported through 1982 use these two reactor types. Both the BWR and PWR are called *light-water reactors* (LWRs), because their coolant, or heat transfer medium, is ordinary water incorporating the light isotope of hydrogen (atomic weight 1) as compared to heavy hydrogen or deuterium (atomic weight 2).

The BWR is similar to the boiler in a fossil fuel power plant in that the heat of the fuel is used directly to boil the feedwater, producing the steam which drives the turbine which turns the electric generator. Instead of burning fossil fuel, the reactor regulates the fission of the nuclear fuel to produce heat. The BWR is fueled by slightly enriched uranium (for example, 2 to 3 percent ^{235}U isotope) in the form of uranium oxide pellets held in zirconium-alloy tubes in the core. Water is pumped through the core, where it is boiled by the heat produced by the fissioning uranium, producing steam which is piped to the turbine. This process gives the reactor the name *boiling-water reactor*. The relationship of the reactor to the feedwater, the main steam, and the other plant systems is shown in Fig. 1.6.

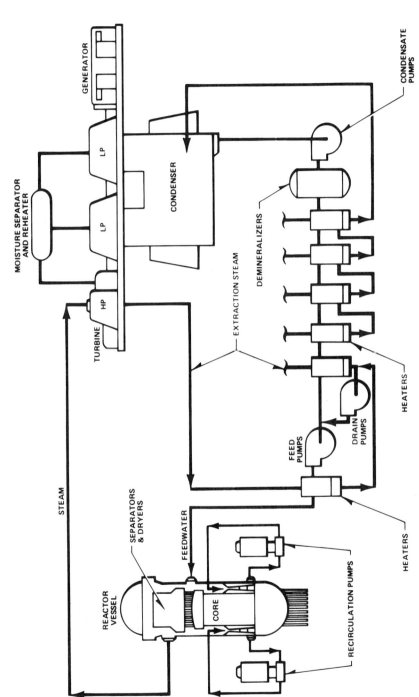

Figure 1.6 Typical boiling-water reactor plant. *(Courtesy of General Electric.)*

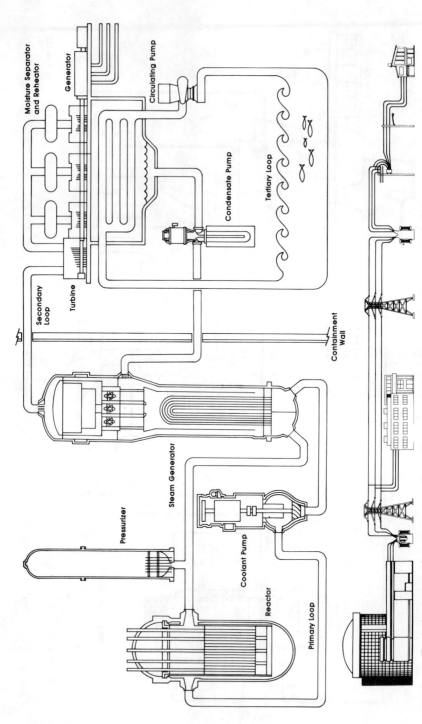

Figure 1.7 Typical pressurized-water reactor plant. *(Courtesy of Westinghouse Electric Corporation.)*

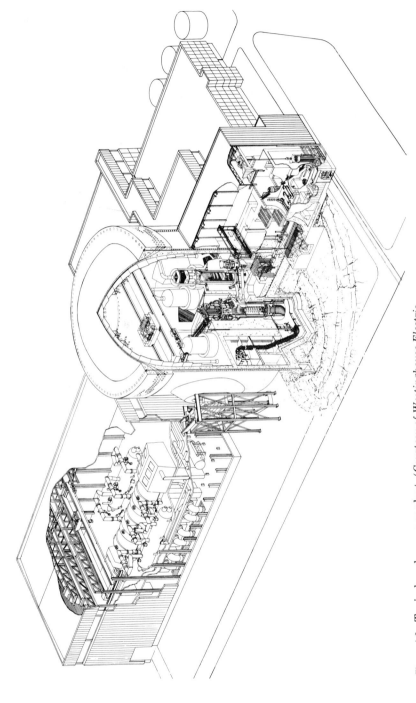

Figure 1.8 Typical nuclear power plant. (*Courtesy of Westinghouse Electric Corporation.*)

The pressurized-water reactor (see Fig. 1.7) is also fueled by enriched uranium oxide pellets held in zirconium-alloy tubes. As in the BWR, water is pumped through the core to transfer heat from the fuel; however, in this case, the water is kept under pressure to prevent boiling, hence the name *pressurized-water reactor*. This coolant is circulated through the tubes of a large exchanger known as the *steam generator,* where heat is transferred to the feedwater on the shell side. The feedwater boils in the steam generator to produce steam, which again is piped to the turbine. This type of reactor has the advantage of maintaining isolation of the reactor coolant (primary side) from the feedwater and main steam (secondary side).

At both types of plants, steam is used to generate electricity in exactly the same way as in a fossil fuel plant. This steam drives the turbine which in turn spins the generator. This equipment is essentially the same in all thermal power plants. The end product, electricity, is identical, too. A nuclear plant, then, is nothing more than a steam-electric generation station in which the reactor takes the place of a furnace and the heat comes from the fissioning of uranium fuel, rather than from the burning of fossil fuel.

The plant illustrated in Fig. 1.8 indicates the tremendous amount of piping that may be required in a power plant as well as the complexity of the geometry and the intricate piping arrangements. Figure 1.8 illustrates the role of piping systems in connecting the major components of the plant, much as veins and arteries carry essential fluids to and from the major organs of the body.

Chapter

2

Codes, Standards, and Regulations

2.1 Introduction

Continuing demands on the power industry to generate economical, safe, and efficient energy has brought forth ever more sophisticated plants and piping systems. Regulation of piping systems has increased for power plants in general and specifically for nuclear power plants.

Standardization can, and does, reduce the cost, inconvenience, and confusion that result from unnecessary and undesirable differences in equipment, systems, materials, and procedures. Standards can also document accepted industry practice in areas such as safety, testing, and installation. Within an engineering organization, standardization is furthered through the vehicle of official company standards and procedures. Between organizations, it is often accomplished through industry standards. Between nations, it is accomplished through international agreements and standards. For example, the General Agreement on Tariffs and Trade (GATT), which became effective January 1, 1980, includes the "Agreement on Technical Barriers to Trade," which essentially establishes a code to promote the use of international standards and to prevent local and national standards, testing, and certification procedures from becoming technical barriers to trade.

Most of the major architectural or engineering consulting firms in the United States and western Europe have developed a number of standard reference power plants, of varying fuel type and output capacity. These standard reference plants reduce the cost allocated to design and engineering, since many components and systems are preengineered and may be interchanged for similar types of plants.

Industry standards are generally published by professional societies, trade organizations, and committees. Standards are adopted by industry consensus after preparation by experts in the field.

To protect industrial employees and the general public, often municipal, state, and federal governments incorporate codes into their laws and regulations. A code is basically a standard that has been legally accepted by a government agency. A code is written with the expectation that it will be adopted as law. The objective of each code is to ensure public and industrial safety in a particular technical activity. Codes are often developed by the same organizations that develop standards. For example, the American Society of Mechanical Engineers (ASME) has an active standards program and has also developed the ASME Boiler and Pressure Vessel Code.

In the nuclear power industry, there is a compelling need for standards to protect the health and safety of the public. Most of the approved nuclear safety standards, and those now in preparation, are related to plant safety.

A brief history of codes and standards applicable to piping systems is presented here along with a discussion of some of the organizations that produce these standards.

2.2 History of Applicable Documents (Codes, Standards, Recommended Practices, and Guides)

The growing use of boilers in the late 1700s led to the need for regulations which would safeguard the public from catastrophic boiler failures. Boilers generate steam with pressures far above that of the atmosphere. Carelessness and negligence of operators, faulty design of safety valves, and inadequate inspections led to many boiler ruptures and explosions in the United States and Europe until sufficient regulations were imposed.

In June 1817, a joint committee of the Council of the City of Philadelphia addressed the subject of steamship boiler explosions. This committee recommended that state legislators institute a law requiring the pressure rating to be listed on the boiler, properly placed safety valves to be used, and monthly inspections.

In the United States, there appears to have been no official regulatory organization in this field until 1866. At that time, the Hartford Steam Boiler Inspection and Insurance Co. was founded for such purposes under a charter granted by the Connecticut legislature.

Most of the earlier explosions resulted in the loss of life of one or two persons and a number of injuries. More disastrous boiler explosions have occurred on occasion, such as on the steamboat *Sultana* on the Mississippi River on April 27, 1865. This accident resulted in the loss of 1500 lives. More than two decades later, the simultaneous explosion of 27 boilers at the Henry Clay Mine in Shamokin, Pennsylvania, leveled the surrounding mining town and killed thousands.

Just over 10 years later, on March 20, 1905, a boiler explosion occurred in a shoe factory in Brockton, Massachusetts, killing 58 persons and injuring 117 others. As a result of the 1905 disaster, Massachusetts adopted the first known legal code of rules for construction of boilers in 1907. In 1908, Ohio adopted rules similar to those of Massachusetts. Other states and municipalities began to adopt legislation for boiler construction, many unique to the legislative body. In 1911, appeals were issued to the council of the ASME to resolve conflicts and prevent future disasters. This led to the ultimate creation of the ASME Code, described later in this chapter.

Boiler explosions demonstrated a need for design standards to protect the public. In addition to the need for boiler regulation for safety purposes, the industrial revolution created the need for standardization of mechanical parts for the smooth operation of assembly lines, the maintenance of plants, and the resolution of disputes.

As a result of the codification of safety requirements for boilers, there has been a remarkable history of improvements in safety. The ASME and Hartford Steam Boiler Inspection and Insurance Co. have data that demonstrates improvements in safety.

In response to obvious design and standardization needs, numerous societies were formed between 1911 and 1921, such as the American Standards Association (ASA), now the American National Standards Institute (ANSI); the American Institute of Steel Construction (AISC); and the American Welding Society (AWS). Codes and standards were established to provide methods for manufacturing, listing, and reporting design data.

The intent of various applicable documents is often misinterpreted, since definitions of a code, standard, recommended practice, or guide are not always understood. The following definitions are generally accepted:

Code. A group of general rules or systematic standards for design, materials, fabrication, installation, and inspection prepared in such a manner that it can be adopted by a legal jurisdiction and made into law.

Standards. Documents prepared by a professional group having requirements believed to be good and proper engineering practice and which are written with mandatory requirements (i.e., the verb *shall* is used).

Recommended practices. Documents prepared by a professional group indicating good engineering practices but which are optional (i.e., the verb *should* is used).

Guides and guidelines. Documents by agencies, organizations, and committees enumerating various engineering methods which are considered good practice, without any specific recommendation or requirement. These guides may be used at the engineer's discretion.

Companies develop engineering and design guides in order to have consistent in-house design procedures and to avoid having one project differ substantially from others. It is another way to standardize the specific firm's design approach. Most of these guides are existing standards modified or adopted to agree with the company's specific design philosophy.

Applicable documents (codes, standards, recommended practices, and guides) are written by committees consisting of persons representing industries, professional societies, manufacturers, consultants, and occasionally regulatory bodies. These codes are usually developed by volunteers who, as individuals, are experts in their fields, providing their accumulated knowledge and experience. Committee members, besides being experts in their fields, must also be of good moral standing since they bear responsibilities as guardians of society's safety.

Applicable documents make known specific areas of concern and identify minimum requirements in the form of basic principles. Generally, codes contain prohibitions or cautions to be followed in the standardization of design practices for safety considerations. Economics and enforceability are basic considerations for any code development and must be weighed against safety considerations. Although some technological lags are unavoidable, codes are reviewed and updated periodically to hold these to a minimum.

Usually the concurrence of an overwhelming majority (between two-thirds and 90 percent) of committee members is required to formulate a code or standard. Codes thus establish design practices acceptable to the engineering profession as a whole. Codes and standards, as well as being regulations, might be considered "design aids" since they provide guidance from experts. If one does not comply with standards, professional liability suits may result, especially if this noncompliance results in an error or omission which causes failure.

Code changes come in the form of addenda, errata sheets, or new editions of the code (generally on a triannual basis). When a code requirement is questioned and an official interpretation is requested, the resulting committee response may relax that requirement; in that case the response would be labeled a *code case* for the ASME codes and be assigned a number for subsequent identification. These code cases do not constitute a revision of the code, and users of the code cases are cautioned to make sure they apply to the given problem. Code changes are usually not retroactive and are mandatory 6 months after the date of issue, unless stated otherwise.

The applicable codes for mechanical equipment are those in effect at the time the equipment purchase order is awarded. These codes and standards should be listed in the system design description, which may have to be updated or revised once the purchase order is placed. Later revisions are usually not required to be implemented unless a safety issue is affected.

If a change occurs in the code and the engineering or construction is under way, the project or design engineer generally brings this code change to the attention of the owner to determine whether compliance is required. The inclusion of the code change to the design basis must be documented.

The owner of a piping system has the ultimate responsibility for compliance with codes, standards, and regulations. The owner must ensure that requirements are established and complied with for the design, construction, examination, inspection, and testing of the piping systems.

2.3 Overview of Applicable Documents Relating to Piping

To protect the public, numerous international, federal, state, and local government regulations have been enacted. In response to these regulations, and in the interest of public safety, various professional societies have published design guidelines, codes, standards, etc. The myriad regulations, standards, and codes in existence create the need to organize available data as a reference to piping engineers in various applications. The following identification of organizations active in the development of piping standards enables the reader to focus on that organization which affects the engineering and design work of a specific industry. This compilation should be used only as a guide and for instructive purposes. The user of a reference cannot substitute this information for the client's specification or for any regulatory commitments.

2.3.1 American Institute of Steel Construction

The AISC was founded in 1921. The first AISC manual was issued in 1926. Presently in its eighth edition, the manual has proved invaluable in providing guidelines and codes to maximize structural steel design efficiency and safety.

The AISC code contains design equations, design criteria, and accepted design practice for structural steel. Its use is recommended for the design of buildings, bridges, or any steel structure, including those serving as pipe restraints.

2.3.2 American National Standards Institute

Initially established in 1918 as the American Standards Association (ASA), this group changed its name in 1967 to USA Standards Institute (USASI). In 1969, the name was changed again to the American National Standards Institute (ANSI). Standards issued are prefixed with ANSI, although they may have originally been termed ASA or USASI.

ANSI is the clearinghouse for more than 200 major organizations which develop standards. It aids in ensuring the precision and quality of the U.S. voluntary standards system and provides the only recognized mechanism in the United States for establishing standards as U.S. national standards. This mechanism requires that every proposed standard submitted to ANSI for approval be subjected to a period of public review and comment. During this period, any interested party may obtain a copy of the draft and submit comments. All comments must be fully considered by the originating committee. ANSI's Board of Standards Review (BSR) then determines, on the basis of the evidence presented, if national consensus exists.

Through its federated membership, boards, councils and committees, ANSI

- Identifies what standards are required and when they are required
- Arranges for competent and willing organizations to undertake their development; or, if none exists, organizes the required competence
- Provides effective procedures that standards-writing organizations can use to ensure timely and effective development
- Supplies management services to ensure efficient use of resources and to eliminate duplication of effort
- Lends its prestige to the standard writers

Not all U.S. standards are issued directly by ANSI. The ASME, the AISC, the AWS, and numerous other organizations issue standards and codes applicable to piping. Nuclear-related standards are published as ANSI N series standards and are authored by standing committees of both the American Nuclear Society (ANS) and the ASME. A listing of all standards and specifications applicable to piping design, together with their mandatory effective edition reference, appears in an appendix of the Code for Pressure Piping (ANSI B31).

The ANSI/ASME B31 Code for Pressure Piping is at present a nonmandatory code in the United States, although most U.S. state legislative bodies and Canadian provinces have adopted these codes as legal requirements. The minimum design requirements of these codes have been accepted by the industry as a standard for all piping outside the jurisdiction of other codes, i.e., the ASME Boiler and Pressure Vessel Code. The piping industries covered by the separate sections of the code are listed in

TABLE 2.1 ANSI B31 Committees

Industry	Description
Power piping	B31.1
Fuel gas piping	B31.2
Chemical plant and petroleum refinery piping	B31.3
Liquid-petroleum transportation piping system	B31.4
Refrigeration piping	B31.5
Nuclear power piping (superseded by ASME, Section III)	B31.7
Gas transmission and distribution piping	B31.8
Building services piping	B31.9

Table 2.1. ANSI also has dimensional standards for pipes and tubes, flanges, bolts, threads, and valves. Since the issuance of standards is an ongoing process, users should find that date of issuance to which they are committed before using the standard in any design.

2.3.3 Manufacturers Standardization Society of the valve and fittings industry

The Manufacturers Standardization Society (MSS) is a national volunteer organization of the valve and fittings industry. The MSS issues advisory standards which are considered standard practice throughout the industry. These MSS standards are applicable to pipe restraints:

SP-58	Pipe Hangers and Supports: Materials, Design, and Manufacturers
SP-69	Pipe Hangers and Supports: Selection and Application
SP-77	Guidelines for Pipe Support Contractual Relationships
SP-89	Pipe Hangers and Supports: Fabrication and Installation Practices
SP-90	Guidelines on Terminology of Pipe Hangers and Supports

2.3.4 American Society of Mechanical Engineers

In 1911, in New York City, a group of engineers met to review laws regarding the construction and operation of steam boilers passed by the Commonwealth of Massachusetts and the Ohio state legislature, subsequent to the 1905 Brockton, Massachusetts, boiler explosion. This meeting was the basis for the formation of ASME. Meetings were held to study the states' codes and to formulate a materials standard. In 1913, the committee issued the first preliminary report to 2000 mechanical engineers, professionals, and insurance inspectors.

Initially issued in 1914, Section I of the ASME code was one of the first codes and standards in the United States. The ASME Code Committee expanded its areas of concern in 1924 by providing design data in Section II which defined specifications for various types of materials. Today, the code covers the design and construction of power boilers, heating boilers, nuclear plant components, and any pressure vessel which operates at a pressure of at least 15 psi. These sections of the code, along with their original issue year, are summarized:

ASME Boiler and Pressure Vessel Code sections

Section I: Power boilers (issued in 1914). This construction code covers power boilers, electric and miniature boilers, and high-temperature boilers used in stationary service. The section includes power boilers used in locomotive, portable, and traction service.

Section II: Material specifications (issued in 1924). This section is divided into three parts (A, B, and C) and contains specifications on code-accepted ferrous materials; nonferrous materials; and welding rods, electrodes, and filler metals.

Section III: Nuclear power plant components (issued in 1963). Division 1 of Section III is subdivided into seven subsections. Division 2 of Section III covers the design of concrete containment vessels. The subsections of Division 1 are as follows:

Subsection	General Description
NA	Rules and design criteria for items which are common to all subsections, thus indicating general requirements.
NB	Rules for Class 1 components, which are those within the reactor coolant pressure boundary. Class 1 components include the reactor vessel, piping, pumps, and valves.
NC	Rules for Class 2 components, which are those that are important to safety and designed for emergency core cooling, accident mitigation, containment heat removal, postaccident fission product removal, and containment isolation. Class 2 components include pressure vessels, piping, pumps, valves, and storage tanks.
ND	Rules for Class 3 components, which are those found in the cooling water and auxiliary feedwater systems. Class 3 components include pressure vessels, piping, pumps, valves, and storage tanks.
NE	Rules for containment vessels, which are referred to as Class MC (metal containment). This subsection contains rules for only the containment vessel and does not include piping, pumps or valves, since those portions of the containment system are generally designed as Class 2.
NF	Subsection NF contains the rules for Class 1, 2, 3, and MC component supports. These may be plate and shell, linear, or standard supports. These items are used to support the vessels, pumps, and piping systems.

Section IV: Heating boilers (issued in 1923). Section IV is a construction code covering the design, fabrication, installation, and inspection of steam heating, hot water heating, and hot water supply boilers which are directly fired by oil, gas, electricity, or coal.

Section V: Nondestructive examination (issued in 1971). This section presents methods accepted for use by the code: radiography, ultrasonics, liquid penetrant, magnetic particles, eddy current, visual and leak testing.

Section VI: Recommended rules for care and operation of heating boilers (issued in 1926). Section VI regulates steel and cast-iron heating boilers, providing a guide to the operation, maintenance, and repair of this equipment. Boiler room accessories and facilities, automatic fuel burning equipment and controls, and water treatments are covered.

Section VII: Recommended rules for care of power boilers (issued in 1922). Similar to Section VI, this section addresses the operation and maintenance of stationary, portable, and traction types of power boilers.

Section VIII: Pressure vessels, Division 1 (issued in 1925). This section provides basic rules for the construction, design, fabrication, inspection, and certification of pressure vessels. These rules are formulated on the basis of design principles and construction practices applicable to vessels designed for pressures up to 8000 psi (55×10^6 Pa). Stamping and coding are also covered.

Section VIII: Pressure vessels, Division 2 (issued in 1925). Division 2 provides an alternative to construction requirements for pressure vessels outlined in Division 1. The rules are more restrictive in the choice of materials which may be used, but the rules permit higher design stress intensity values in the range of temperatures over which the design stress intensity value is controlled by the ultimate or yield strength. More precise design procedures are required, and some common design details are prohibited. Permissible fabrication procedures are delineated, and more complete testing and inspection are called for. The rules in this section apply to vessels to be installed at stationary locations.

Section IX: Welding and brazing qualification (issued in 1937). This section covers the qualification of welders and welding procedures in order to comply with the code. Under procedure qualifications, each process is listed and the essential and nonessential variables of each process are spelled out. Welding performance qualifications are also included.

Section X: Fiberglass-reinforced plastic pressure vessels (issued in 1969). This section establishes general specifications for the glass and resin used to fabricate such vessels. It sets limits on the permissible service conditions and sets rules under which fabricating procedures are qualified. It also outlines the requirements for stamping and marking.

TABLE 2.2 ASME Code Stamps and Their Coverage

A	Field assembly of power boilers and of steel-plate heating boilers
H	Steel-plate and cast-iron sectional heating boilers
HLW	Lined portable water heaters
L	Locomotive boilers
M	Miniature boilers
N	Nuclear vessels and piping systems
NPT	Nuclear vessel parts
NA	Nuclear installation
NV	Nuclear vessel safety valves
PP	Pressure piping
RP	Reinforced-plastic pressure vessels
S	Power boilers
U	Pressure vessels (Division 1)
U2	Pressure vessels (Division 2)
UM	Miniature pressure vessels
UV	Pressure-vessel safety valves
V	Boiler safety valves

Section XI: Rules for in-service inspection of nuclear power plant components (issued in 1970). This section outlines the requirements for maintenance of the nuclear power plant in a safe operating condition and for returning the plant to service following a refueling or maintenance shutdown.

The ASME code, including all the sections previously discussed, is mandatory in most states, and compliance with the code is required in the United States and Canada in order to qualify for certain insurance and operating licenses. Additionally, foreign countries must comply with the code in order to receive nuclear fuel from the United States.

Since the code has numerous editions, it is important that the code year edition, as well as the applicable addenda with which the project design must comply, be expressly stated in contract documents.

The ASME issues fabricators of equipment within the jurisdiction of the code official stamps certifying their engineering and quality control competence. The ASME code is the only code known to require third-party independent inspection. Failure to satisfactorily meet inspection requirements will cause the ASME to force the fabricator to cease this type of fabrication. The ASME code stamps and their coverage are listed in Table 2.2 giving an indication of the areas addressed by the codes.

Discussion of Uniform Boiler and Pressure Vessel Laws Society and the National Board of Boiler and Pressure Vessel Inspectors later in this chapter provides an overview of the legal aspects of the code. Note that the ASME code is typically utilized even when the local jurisdiction does not make it mandatory. Insurance company requirements or legal requirements at the place of either fabrication or ultimate use usually make it unlikely that a noncode component would be used. Within the United

States, Title 10 of the Code of Federal Regulations (Part 40.55a, "Codes and Standards"), makes ASME code compliance mandatory for many nuclear power plant components. These requirements are binding regardless of the local jurisdiction's nonadoption of the code.

2.3.5 Jurisdictional boundaries

When more than one code provides criteria for a system or components, or when various sections of a code give different criteria, the limit of those criteria must be defined. It is generally understood that building structures are designed according to the AISC code, and piping and other mechanical components are designed according to the ASME code. The transition point between the building structure and the pipe support structure determines which code is applicable. Figures 2.1 and 2.2 mark the particular code group having jurisdiction. The owner usually provides a design specification which defines these jurisdictional boundaries.

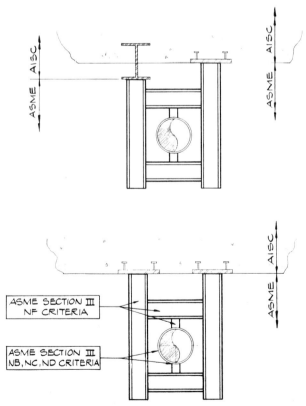

Figure 2.1 Authors' interpretation of ASME and AISC jurisdictional boundaries between building structure and pipe support structure.

28   Piping and Pipe Support Systems

The current issues of the AISC and ASME codes have not sanctioned these boundary definitions. Therefore, contractors are cautioned to have jurisdictional boundaries defined for their specific projects. The engineer shall specify in the technical specifications the boundaries of building structures. Figure 2.3 shows the jurisdictional boundaries between different classes of piping. Note that in the figure the piping class changes at a valve in the pipeline, with the valve falling within the jurisdiction of the more restrictive piping class.

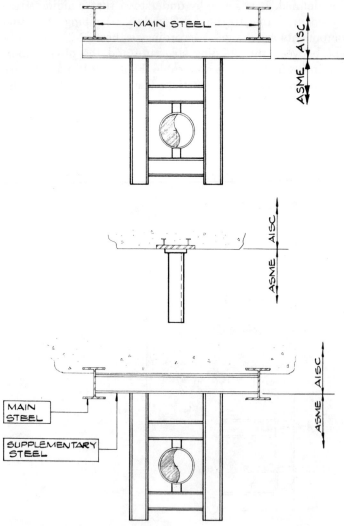

Figure 2.2 Authors' interpretation of complex ASME and AISC boundaries for pipe supports in which supplementary building steel is required.

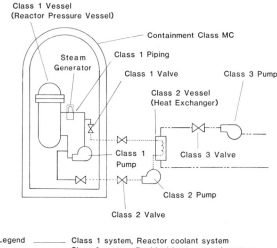

Figure 2.3 Component classification—typical code jurisdiction in a PWR power plant.

2.3.6 American Society for Testing and Materials

The American Society for Testing and Materials (ASTM) was founded in 1898 to develop standards on the characteristics and performance of materials, products, suppliers, and services and to promote related knowledge. The 1983 ASTM standards consist of 66 volumes divided among 16 sections on standard specifications, test methods, classifications, definitions, practices, and related topics. All acceptable piping materials have been tested and assigned ASTM specification numbers and grades.

Based on ASTM testing and specifications, the ASME and AISC codes establish the maximum allowable stresses for materials as a function of temperature. Usually ASME-accepted pressure-retaining materials are identified by an S preceding the A specified by the ASTM; for example, A-36 is an ASTM-endorsed structural steel specification while SA-36 is an ASME-endorsed piping material.

2.3.7 Uniform Boiler and Pressure Vessel Laws Society

This is a nonpolitical, noncommercial, nonprofit, technical body whose objective is to secure uniformity in the laws, rules, and regulations which affect the boiler and pressure-vessel industry, inspection agencies, and users. The society promotes nationally accepted codes and standards and recommends the ASME Boiler and Pressure Vessel Code as the standard for construction and the Inspection Code of the National Board of Boiler and Pressure Vessel Inspectors for administration and inspection.

Although several states have not adopted the ASME code, the code as it relates to nuclear plants is in effect since the Nuclear Regulatory Commission (NRC), in its Code of Federal Regulations 10CFR50.55(a), mandates that the ASME code (Section III) is required. In addition, the use of the ASME code is required for insurance purposes.

2.3.8 National Board of Boiler and Pressure Vessel Inspectors

This group promotes (1) uniform enforcement of boiler and pressure-vessel laws, rules, and regulations; (2) standards for acceptance of boilers, pressure vessels, and parts; (3) a standard of qualifications and examinations for inspectors who are to enforce the requirements of the code; and (4) the use of one uniform code and one standard stamp to be placed on all registered boilers, pressure vessels, and parts constructed in accordance with the code. The National Board of Boiler and Pressure Vessel Inspectors is the enforcement agency of the ASME Boiler and Pressure Vessel Code.

2.3.9 American Society of Heating, Refrigeration, and Air Conditioning Engineers

The purpose of the American Society of Heating, Refrigeration, and Air Conditioning Engineers (ASHRAE) is to advance the technology of heating, refrigeration, air conditioning, and ventilation for the benefit of the general public.

2.3.10 Pipe Fabrication Institute

The purpose of the Pipe Fabrication Institute (PFI) is to serve proven needs of the pipe fabrication industry both at the design level and in actual shop operations, by providing recommended procedures which have been demonstrated by collective experience to fulfill code compliance requirements.

2.3.11 Occupational Safety and Health Administration

The Occupational Safety and Health Administration was created by Congress; therefore its regulations hold the force of law in all states. OSHA addresses such matters as clearance for means of access and egress and maintaining sufficient headroom in work environments. For example, according to OSHA rules, the ceiling height shall not be less than 7 ft 6 in, and no piping components shall project lower than 6 ft 8 in above the floor. OSHA requirements must be carefully reviewed by piping designers to determine other items that may affect design.

2.3.12 American Welding Society

The AWS was founded in 1919 and is the national organization for advancing the art and practice of welding. The AWS provides information on welding fundamentals, weld design, welders' training qualification, testing and inspection of welds, and guidance on the application and use of welds.

2.3.13 Building codes

Building codes include requirements relating to the structural soundness and safety of buildings. Within the United States most states have adopted one of three national codes (some with state amendments). The Standard Building Code (SBC), developed by the Southern Building Code Congress International (SBCCI), is generally used in the southeast. The Basic Building Code (BBC), developed by Building Officials and Code Administrators International Inc. (BOCA), is generally used in the northeast and midwest. The Uniform Building Code (UBC), developed by the International Conference of Building Officials (ICBO), is generally used in the west.

2.3.14 Nuclear Regulatory Commission

The Atomic Energy Commission (AEC), which was the forerunner of the NRC, was founded in 1946 as a result of an act of Congress. The AEC was abolished and the NRC was created under the Energy Reorganization Act of 1974. During 1975, responsibilities were further divided between the NRC and the Energy Research and Development Administration (ERDA), separating the regulatory and developmental aspects of nuclear power. All phases of the nuclear power industry are regulated by the NRC. The development of energy forms, such as solar, geothermal, nuclear, etc., is governed by the Department of Energy, which has succeeded the ERDA.

The NRC is a regulatory agency which acts as the "enforcement arm" of the federal government, issuing nuclear requirements and regulations and, above all, construction permits and operating licenses for nuclear plants. The NRC's rules and regulations are issued through the Code of Federal Regulations (CFR).

License applicants and licensees of nuclear plants must satisfy all NRC regulations covering radiation protection, nuclear materials, site criteria, and physical security. The following parts of the CFR are of particular concern for nuclear-related piping systems:

10CFR 21   Reporting of Defects and Noncompliance
10CFR 50   Licensing of Production and Utilization Facilities
10CFR 100   Reactor Site Criteria

The NRC issues numerous documents in order to give guidance on or warning of deficiencies, to provide data on safety-related research, and to inform industry, governments, and the public about potential problems. These documents are discussed in the following sections.

2.3.14.1 Regulatory guides. The NRC issues regulatory guides to describe and make available to the public methods acceptable to the NRC staff for implementing specific parts of the NRC regulations, to delineate techniques used by the staff in evaluating specific problems or postulated accidents, and to provide guidance to applicants. Regulatory guides are not substitutes for regulations, and compliance with them is not always legally required. Methods and solutions different from those set out in these guides are acceptable if approved by the NRC. Regulatory guides that have a major impact on piping design are listed in Table 2.3.

The regulatory guides are generally divided into four sections:

1. *Introduction.* This part describes regulation requirements and an acceptable method of compliance.
2. *Discussion.* This part describes the general background with an explanation as to how the method of compliance was established.
3. *Regulatory position.* This section describes the acceptable method of compliance in detail and discusses the applicability, if any, of ANSI standards to the method of compliance.
4. *Implementation.* This phase is required only after the applicant commits to a suggested method of compliance.

Although applicants have the alternative of developing their own methods of compliance, it is usually not feasible to do so because of the lengthy and complex procedures involved. Levels of documentation include the safety analysis report (SAR), quality assurance manual, implementing procedures, operating controls, and specific instructions.

2.3.14.2 NRC inspection and enforcement directives. When the NRC becomes aware of inadequacies or defects in a facility that seem to be of a generic nature, the NRC Office of Inspection and Enforcement may

Codes, Standards, and Regulations 33

TABLE 2.3 NRC Regulatory Guides Having a Major Impact on Piping Design

Number	Title
1.26	Quality Group Classifications and Standards for Water, Steam, and Radioactive-Waste-Containing Components of Nuclear Power Plants (Revisions 3, 2/76)
1.28	Quality Assurance Program Requirements (Design and Construction) (Safety Guide 28, 6/7/72)
1.29	Seismic Design Classification (Revision 2, 2/76)
1.46	Protection against Pipe Whip Inside Containment (5/73)
1.48	Design Limits and Loading Combinations for Seismic Category I Fluid System Components (5/73)
1.57	Design Limits and Loading Combinations for Metal Primary Reactor Containment System Components (6/73)
1.60	Design Response Spectra for Seismic Design of Nuclear Power Plants (Revision, 1, 12/73)
1.61	Damping Values for Seismic Design of Nuclear Power Plants (10/73)
1.64	Quality Assurance Requirements for the Design of Nuclear Power Plants (Revision 2, 6/76)
1.67	Installation of Overpressure Protection Devices (10/73)
1.68	Initial Test Programs for Water-Cooled Reactor Power Plants (Revision 1, 1/77)
1.68.1	Preoperational and Initial Startup Testing of Feedwater and Condensate Systems for Boiling Water Reactor Power Plants (Revision 1, 1/77)
1.84	Code Case Acceptability—ASME Section III Design and Fabrication (Revision 10, 1977)
1.85	Code Case Acceptability—ASME Section III Materials (Revision 10, 1977)
1.88	Collection, Storage, and Maintenance of Nuclear Power Plant Quality Assurance Records (Revision 2, 10/76)
1.92	Combining Modal Responses and Spatial Components in Seismic Response Analysis (Revision 1, 2/76)
1.96	Design of Main Steam Isolation Valve Leakage Control Systems for Boiling Water Reactor Nuclear Power Plants (Revision 1, 6/76)
1.122	Development of Floor Design Response Spectra for Seismic Design of Floor-Supported Equipment or Components (Revision 1, 2/78)
1.124	Service Limits and Loading Combinations for Class 1 Linear-Type Component Supports (Revision 1, 1/78)
1.130	Service Limits and Loading Combinations for Class 1 Plate and Shell-Type Component Supports (Revision 1, 10/78)
1.144	Auditing of Quality Assurance Programs for Nuclear Power Plants (Revision 1, 9/80, 2.3.21)

issue a bulletin to a group of licensees requiring inspection, reporting, and correction of the situation, as appropriate. The bulletin will specify the action required of each class of licensee and the schedule for completion of the action. Also, the bulletin will specify reporting requirements (in some cases, reporting is not required, and inspectors will review compliance at the individual facility). The NRC may modify a license if the holder fails to respond to the action requirements of a bulletin. Thus, a bulletin carries legal force. Rather significant design and evaluation requirements for nuclear plant piping and restraints are included in these bulletins. For example, IE Bulletin 79-14, Seismic Analyses for As-Built Safety-Related Piping Systems, which required operating nuclear plants to reevaluate the seismic resistance of their piping systems, caused large amounts of reanalysis and design, resulting in construction changes in many cases.

Occasionally the NRC is informed about a situation which does not meet the criteria for a bulletin in either significance or immediacy, but could have potential generic implications. In these cases an inspection and enforcement circular is issued. Generally, a circular *recommends* that licensees review the information provided and take certain specific preventive actions. Since the circular identifies potential problems not of immediate health and safety significance, no reporting is required. Thus, the circular is basically a mechanism for the dissemination of information on problem areas and recommended licensee actions.

The newest category of inspection and enforcement (IE) issuance is the *information notice*. The IE information notice provides early notification of a *possibly* significant matter. It is issued when the NRC becomes aware of a potentially significant matter but does not yet know the extent or seriousness of the deficiency. Licensees are expected to review the information for applicability to their facility, but no specific action is required and no response is required. If further evaluation warrants additional licensee action, the requirement will be upgraded by issuance of a circular or bulletin.

Although circulars and information notices require no formal response, the Office of Inspection and Enforcement has indicated that licensee review of these documents and resultant actions are subject to audit by their inspectors.

2.3.14.3 NUREG reports. These documents are formal reports issued by the NRC which carry the NUREG series code. They may contain information to be considered in NRC licensing and policy decisions, may present the results of NRC licensing and policy decisions, or may present the results of technical research conducted for the NRC. There are four categories of NUREG reports:

- NUREG: technical or procedural reports written by one or more NRC offices
- NUREG/CR: technical reports written as a product of NRC-sponsored research (i.e., contract report)
- NUREG/CP: documentation of discussions held at NRC-sponsored meetings, regional meetings, or workshops (i.e., conference proceedings)
- NUREG/TR: technical reports translated for the NRC

Generally, NUREG reports have no official standing as requirements and in most cases serve only to document a current situation. In those cases where a NUREG report is written to document a staff position, it must be imposed by some other vehicle such as an IE bulletin.

2.3.14.4 Licensee event reports. For each licensed facility, certain abnormal events must be reported to the NRC. Reporting requirements for licensed facilities are defined in 10CFR50, in license provisions, and in the plant unique technical specifications. The abnormal events are designated "reportable occurrences," and the reports are known as licensee event reports (LERs).

To collect, sort, store, retrieve, and evaluate information concerning licensee events promptly, the AEC established, in the summer of 1973, a computer-based LER file. This file provides a centralized source of data for assessment of abnormal events in the nuclear industry as well as a reference and index to the source material (LER) for more detail. This file is accessible not only to the NRC but also to the industry and public.

2.3.14.5 Standard review plans. The director of licensing of the NRC has prepared regulatory standard review plans (SRPs) as guidance for use in the review of applications to construct and operate nuclear power plants. There are currently over 200 SRPs. Each SRP is identified by the safety analysis report (SAR) section to which it applies. The SRPs specify what technical information is required in order to obtain a construction permit or an operating license.

SRPs are prepared for the guidance of the staff reviewers of the Nuclear Reactor Regulation (NRR) office of the NRC who perform the detailed safety review of applications to construct or operate nuclear power plants. A primary purpose of the SRP is to improve the quality and uniformity of staff reviews. A second purpose is to present a well-defined base from which to evaluate proposed changes in the scope and requirements of reviews. Another purpose of the SRP is to implement NRR policy by making information about regulatory matters widely available. This is intended to improve understanding of the staff review process by interested members of the public and the nuclear power industry.

Table 2.4 lists the SRPs which have a major impact on the design of piping and restraints.

2.3.15 Technical publications

Technical papers may be prepared by industry groups, educational organizations, private individuals, or engineering firms. These papers may be issued directly or published in trade journals. Examples of useful technical papers are the Welding Research Council (WRC) Bulletin 198 ("Secondary Stress Indices for Integral Structural Attachments to Straight

TABLE 2.4 NRC Standard Review Plans Having a Major Impact on Piping Design

Number	Title
3.2.1	Seismic Classification
3.2.2	System Quality Group Classification
3.6.1	Plant Design for Protection against Postulated Piping Failures in Fluid Systems outside Containment
3.6.2	Determination of Break Locations and Dynamic Effects Associated with the Postulated Rupture of Piping
3.7.1	Seismic Input
3.7.2	Seismic System Analysis
3.9.2	Dynamic Testing and Analysis of Mechanical Systems and Components
3.9.3	ASME Code Class 1, 2, and 3 Components, Component Supports, and Core Support Structures
3.9.4	Control Rod Drive Systems
3.9.6	In-service Testing of Pumps and Valves
6.3	Emergency Core Cooling System Performance Requirements

Pipe" and "Stress Indices at Lug Supports on Piping Systems") and 107 ("Local Stress in Spherical and Cylindrical Shells due to External Loadings").

2.3.16 International standards

Other countries also issue standards and codes. For example, the Deutsches Institut Für Normung (DIN) standard is used in West Germany. The British Standards Institute (BSI) in the United Kingdom also issues many standards.

Internationally, there are two nongovernmental groups whose sole purpose is the coordination and approval of voluntary international standards: the International Standardization Organization (ISO) and its counterpart in the electrical/electronics fields, the International Electrotechnical Commission (IEC). ISO recognizes ANSI as the official U.S. member, and IEC accords similar recognition to the U.S. National Committee on IEC, which is affiliated with ANSI.

Recently, a third body, the Pacific Area Standards Congress (PASC), was formed by ANSI in cooperation with the national standards associations of Canada, Japan, Australia, and New Zealand. The purpose of the PASC is to strengthen ISO and IEC and the ability of the nations of the Pacific rim to participate in these organizations.

2.3.17 International Atomic Energy Agency

The United States participates in the drafting, reviewing, amending, and approving of International Atomic Energy Agency (IAEA) documents through the office of NRC. Note that compliance with the Codes of Practice and Safety Guides issued by the IAEA is required only for member states who enter into an agreement with the IAEA for the agency's assistance in connection with the siting, construction, commissioning, operation, or decommissioning of a nuclear power plant.

2.3.18 Atomic Industrial Forum

The Atomic Industrial Forum (AIF) is a nonprofit association composed of organizations and individuals interested in the development and utilization of commercial nuclear energy. Established in 1953, the Forum today has more than 600 organizational members in the United States and some 20 foreign countries. Through conferences, workshops, publications, and expert committee activities, the AIF keeps the public and nuclear industry abreast of technical and economic issues. The AIF's public affairs and information program staff produce many background publications and publish nontechnical brochures and a regular newsletter, INFO. Also, the monthly magazine Nuclear Industry is published by AIF.

2.3.19 Institute of Nuclear Power Operations

The electric utility industry established the Institute of Nuclear Power Operations (INPO) to ensure high-quality operation of nuclear power plants. Its purposes, in brief, are to establish industrywide benchmarks for excellence in nuclear operation and to conduct independent evaluations to assist utilities in meeting these benchmarks. INPO also determines educational and training requirements for operating personnel and accredits training organizations.

2.3.20 National Fire Protection Association

The National Fire Protection Association (NFPA) deals primarily in safety and protection regulations relating to fire hazards. NRC Regulatory Guide 1.120 adopted many of the NFPA standards, including requirements for piping.

2.3.21 American Nuclear Society

The American Nuclear Society (ANS), established in 1954, is a nonprofit scientific and educational organization made up of approximately 12,000 individual scientists and engineers active in nuclear science and technology. Its main objectives include the advancement of science and engineering, integration of the scientific disciplines, encouragement of research, and dissemination of information, all with respect to nuclear science and technology.

2.4 Role of the Quality Assurance Organization within the Nuclear Industry

The NRC has made it clear that electric utilities and their suppliers must develop documented quality assurance practices applicable to all work associated with the design, construction, and operation of nuclear power plants. Determining the scope of a total quality assurance program is senior management's responsibility. The project's senior management determine the extent of the total quality assurance program for the project.

Within a project, a specific group of people are held responsible for planning, auditing, and controlling quality. They may be called inspection, quality control, quality assurance, etc. Regardless of the group's identification, it must develop quality objectives. Through the setting of quality objectives, the project personnel can concentrate on essentials, develop specific programs to achieve the desired results, have a visible and agreed-upon timetable for their accomplishment, and measure progress.

Each project has its unique requirements. Therefore, a total quality assurance program must be tailored to fit the particular project. The quality program and its supporting systems must be adapted to the project operating policies. A successful quality assurance program should not be blindly adopted by the management of any other organization, although it can serve as an excellent beginning point.

There currently exist four major documents that define exactly what the basis for these practices should be:

- Title 10, Code of Federal Regulations, Part 50, Appendix B
- American National Standards Institute (ANSI) N45.2
- Pressure Vessel Code (ASME), Section III, Subsection NCA-4000
- ANSI/ASME NQA-1

Although these four documents have distinct characteristics which make them appear unique, the bulk of the requirements contained in each are essentially the same. All four standards consist of at least 18 sections or areas of involvement:

1. Organization
2. Program
3. Design control
4. Procurement and procurement document control
5. Instructions, procedures, and drawings
6. Document control
7. Control of purchased material, equipment, and services

8. Identification and control of materials, parts, and components
9. Control of special processes
10. Inspection
11. Test control
12. Control of measuring and test equipment
13. Handling, storage, and shipping
14. Inspection, test, and operating status
15. Control of nonconformances
16. Corrective action
17. Quality assurance records
18. Audits

The quality assurance department is responsible for developing the overall program plan, auditing, and ensuring that weaknesses and deficiencies are identified and corrected. Writing the manual and the procedures is only part of the program. The critical point is implementation of the program. Quality assurance provides those tools for assuring and evaluating quality.

Chapter 3

Technical Piping Documentation

3.1 Introduction

To perform all the engineering and construction activities associated with a project, a "project team" may be organized under the direction of a project manager. The project manager is fully responsible for all activities on a specific project and ensures maximum control and accountability. The various disciplines working on the project submit to the manager on questions of planning, scheduling, and cost, although usually not on technical matters.

In the engineering department, a project engineer is assigned to coordinate the numerous engineering functions. The project engineer in turn selects the engineers who lead the major engineering disciplines. Figure 3.1 depicts a commonly used project organization chart, illustrating the major engineering departments such as mechanical, civil and structural, electrical, instrumentation and controls (I&C), and licensing as well as other smaller engineering groups. The types of engineering disciplines (and the number of engineers assigned to each) will depend on the type and size of the project.

The mechanical engineering department assigns engineers to a specific project in four major areas of responsibility: systems; components; building services and heating, ventilation, and air conditioning (HVAC); and piping systems. The piping engineer is given by the engineering organization the responsibility and authority to manage and coordinate the piping program in a manner that will result in meeting the overall project objectives. These responsibilities include the following specific tasks:

- Piping engineering, design, and layout
- Pipe stress analysis

- Pipe support design
- Pipe rupture restraint and jet shield design (if applicable)
- Coordination of piping fabrication contract

The duties of the piping engineer include interfacing with the other project disciplines to ensure that the piping and associated components are delivered to the site and erected in accordance with the codes and standards, technical specifications, construction schedule, and specified budget.

There are a number of technical documents that the piping engineer helps to develop which provide methods and tools needed to control all phases of design, analysis, procurement, fabrication, and installation of piping, supports, and in-line components comprising the piping systems.

The piping engineer reviews the project requirements to establish which documents are needed and when they are to be submitted to the client and to other project participants for review and approval. The preparation of these documents is normally completed during the three phases of plant design. Typical piping documents include

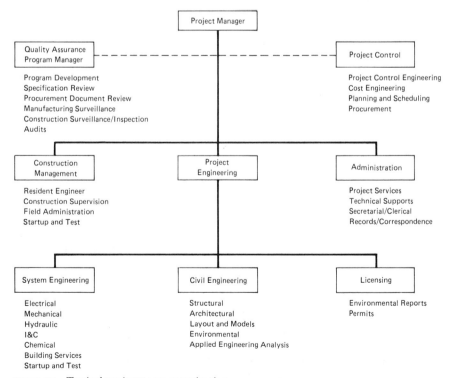

Figure 3.1 Typical project team organization.

Flow diagrams
Piping and instrumentation drawing
System design description
Piping drawings
Stress isometrics
Fabrication isometrics
Support and restraint drawings
Stress analysis reports
Pipe rupture drawings
Jet impingement drawings
Certified spool piece sheets
Erection isometrics
Technical specifications
Design specifications
Valve and miscellaneous in-line drawings
Composite drawings
Scale model
Piping in-service inspection drawings
Engineering and design guides
Construction department schedule
As-built drawings
Design and field changes reports

The three phases of plant design are

- *Phase 1.* Project specification is prepared, and major equipment orders (i.e., boiler, turbine) are placed.
- *Phase 2.* Project estimates and cash flow requirements are prepared. Detail design is begun, with the greatest effort placed on the preparation of specifications, drawings, and equipment lists.
- *Phase 3.* The detail design and drawings are finalized.

This chapter is mainly concerned with the preparation of those documents completed during phase 3 of the project development. In this phase there is a closely coordinated exchange of information among the major disciplines cooperating on the project.

3.2 Flow Diagrams, Line List and Design Specifications

The logical basis for the piping system design, and therefore all piping design drawings, is the flow diagram. This is prepared by the systems engineer and provides the master plan of each *plant system*. It indicates the types of process equipment and interconnecting piping required to perform the function for which the system is intended.

The flow diagram, also known as the *piping and instrumentation drawing* (P&ID), is usually the initial drawing in the piping design process. This document shows the equipment in the system, the instrumentation requirements of the process, and the piping connections between the major equipment. The P&ID shows the required supporting services which impose design requirements on the system.

A sample flow diagram is shown in Fig. 3.2. The flow diagrams are usually representative and not drawn to scale; therefore, emphasis is placed

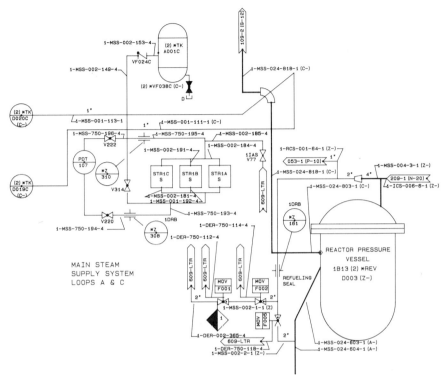

Figure 3.2 Typical flow diagram. *(Courtesy of Construction Systems Associates, Inc., Marietta, Ga.)*

on the schematic relationships between piping and equipment and the design process necessary to satisfy the system design specification.

Each line indicated on the flow diagram is usually identified by a unique line number. This line number is used to reference the subject line's entry in the project piping roster, which is known as the *line list*. The line number remains constant as long as the piping design parameters remain constant. Therefore, when a line number changes, it may be expected that line service, size (diameter or wall thickness), material, temperature, pressure, some other parameter, or any combination thereof, has also changed. This change of line number thus serves as a "flag" to the engineer that a change has occurred.

The pipe line number provides at a glance a minimum of information:

- Pipe nominal diameter
- System to which the pipe belongs
- Unique identifying number for the line
- Pipe safety class (if applicable)

For example,

- 3-BF-8
- 24-MS-51
- 16-CC-2-34

These line numbers refer to a pipe of 3-in (80-mm) nominal diameter in the boiler feed system, a 24-in (600-mm) nominal-diameter main steam line, and a 16-in (400-mm) nominal-diameter containment cooling line (nuclear safety Class 2), with unique identifiers 8, 51, and 34, respectively.

The unique identifying numbers are used, as mentioned, to trace the line to the line list. The line list contains all lines on the project, grouped first by system and then numerically by identifying number. The line list provides all design parameters for the line, including, but not limited to, pipe diameter, wall thickness or schedule, fluid contents, temperature (design and operating), pressure (design and operating), material, insulation thickness, and applicable code. A sample page from a line list is shown in Fig. 3.3.

In addition to a line list, most projects maintain a valve list of all valves used in the piping systems. The valve number, unique for each valve, identifies the system, class, and possibly type of valve. Typical valve numbers are 1-MSS-MOVF-0001 and 1-MSS-V-0220. These numbers identify Class 1 valves from the main steam system. The first valve is a flanged motor-operated valve, while the second is a manually operated valve. The trailing numbers, 0001 and 0220, uniquely identify these valves.

The piping design specification issued for each specific job describes the criteria to be used in the design and construction of the piping systems required for the project. This specification dictates requirements concerning applicable codes, piping materials, fabrication techniques, preferred components, and supports.

Piping materials must be considered in the piping specification, not only because allowable stresses vary with material but also because corrosion is a major concern. *Corrosion* may be defined as the undesired deterioration of a material through chemical or electrochemical interaction with the environment or destruction of materials by pure mechanical action. Failure occurs by corrosion when the corrosive agent renders the piping system incapable of performing its design function.

The complexity of this corrosion process has environmental, electrochemical, and metallurgical aspects. Depending on the combination of environmental, loading, and mechanical factors, numerous types of corrosion mechanisms can be active simultaneously. The rate of corrosion is affected by the following environmental effects:

1. *Temperature.* In nearly all cases, an increase in temperature results in an increase in the corrosion rate.

```
                    NUCLEAR POWER PLANT - UNIT # 1
                       COMPONENT CONTROL SYSTEM
                                                    PAGE :      1
                    P I P E   L I N E   L I S T     DATE :   7/10/84
------------------------------------------------------------------------
                                       WALL   OPERATING    DESIGN
LINE NUMBER              SIZE   CLASS  THCK   PRES   TEMP  PRES   TEMP
                         (IN)          (IN)   (PSIA) (F)   (PSIA) (F)
------------------------------------------------------------------------

1DER-002-365-4           2.000  151    0.191   15.0  212     50    300

1DER-125-118-4           1.250  151    0.075   10.0   60     10    100

1DER-750-112-4           0.750  901    0.042   15.0  212   1250    575

1DER-750-114-4           0.750  901    0.042   15.0  212   1250    575

1MSS-001-111-1           1.000  502    0.157 1022.0  550   1250    575

1ICS-006-006-1           6.000  911    0.352 1205.0  140   1525    575

1MSS-001-112-1           1.000  502    0.157 1022.0  550   1250    575

1MSS-001-192-4           1.000  512    0.157 1000.0  550   1250    575

1MSS-002-001-1           2.000  901    0.191 1050.0  550   1250    575

1MSS-002-002-1           2.000  901    0.191 1050.0  550   1250    575

1MSS-002-072-1           2.000  901    0.191 1050.0  550   1250    575

1MSS-002-149-4           2.000  153    0.012  111.8  120    135    150

1MSS-002-153-4           2.000  153    0.013  111.8  120    135    150

1MSS-002-181-4           2.000  153    0.012  111.8  120    135    150

1MSS-002-184-4           2.000  151    0.011  111.8  120    135    150

1MSS-002-185-4           2.000  151    0.011  111.8  120    135    150

1MSS-002-191-4           2.000  151    0.011  111.8  120    135    150

1MSS-004-003-1           4.000  901    0.181 1050.0  550   1250    575

1MSS-024-603-2          24.00   901    0.968  988.0  540   1250    575

1MSS-024-604-2          24.00   901    0.968  988.0  540   1250    575

1MSS-024-803-2          24.00   901    0.968  988.0  540   1250    575

1MSS-024-818-2          24.00   901    0.968  988.0  540   1250    575

1MSS-750-193-4           0.750  153    0.005  111.8  120    135    150

1MSS-750-194-4           0.750  153    0.005  111.8  120    135    150

1MSS-750-195-4           0.750  153    0.005  111.8  120    135    150
```

Figure 3.3 Line list for pipes.

2. *Velocity.* In general, increasing the velocity at which a fluid passes over a metal surface will tend to increase the corrosion rate.
3. *Fluid composition.* This factor strongly affects the corrosion rate of any material. Factors to consider are the pH of the fluid, the concentration of dissolved solids, and the presence of oxidizing or reducing agents.

Corrosion processes have been classified in many different ways. Several types of corrosion are direct chemical attack, galvanic corrosion, crevice corrosion, pitting corrosion, intergranular corrosion, erosion corrosion, cavitation corrosion, biological corrosion, and stress corrosion.

3.3 Piping Layout

The flow diagrams, line list, and design specification are used by the piping designer to lay out the piping and to generate the design drawings. Piping of the size and schedule indicated must be routed between equipment as shown on the flow diagrams. Routing will be affected by system operating temperature, pipe weight, installation and material costs, applicable code requirements, pressure drop requirements, and equipment and building structure locations. The major equipment and building structure locations should be available on general arrangement and structural drawings.

Pipe routing may vary depending on the design criteria involved. When simple weight loading is involved, the pipe can be easily supported at large distances with a minimal amount of hardware. Thus, for non-safety-related piping, the primary concern is to minimize the pipe length by choosing a more direct route. If the loads are other than simple weight loading, the pipe may need to be close to the building steel. Therefore, for critical piping, the system should be located adjacent to the walls and ceiling, possibly resulting in the use of more piping but also savings in the cost of supports.

Other considerations affect pipe routing:

1. The expansion of the pipe and any attached equipment during operation must be accommodated by the flexibility of the piping run.
2. Pumps, turbines, and other rotating equipment usually require low nozzle reactions. Therefore, the pipe routing in this area should permit the location of pipe supports close enough to the nozzle to take loads off of the equipment.
3. Pipes of the same safety classification should be grouped, if possible, with the pipes resting at the same elevation in order to promote the use of common supports.

4. All high piping points should be vented, and all low points in the system should be drained. Intermediate pockets should be avoided.
5. A minimum headroom of 7 ft (2.1 m), measured from the lowermost point of the piping, should be maintained in pedestrian areas.
6. Piping congestion should be avoided in front of pumps or other equipment requiring frequent maintenance; piping should not be run above pumps.
7. In-service inspection requirements (for nuclear safety-related piping) must be considered in the piping layout.

Accessibility and maintenance requirements are major concerns during piping layout. All piping, components, and supporting structures should be arranged in a neat and simple manner. Designs should allow ample room for free access to all operating equipment, to permit unrestricted operation, component removal, and sufficient maintenance laydown area. Free area should be provided near the location of in-line equipment such as valves, valve operators, snubbers, etc., for the same reasons. Suitable access and maintenance clearances are illustrated in Figs. 3.4 and 3.5. Fol-

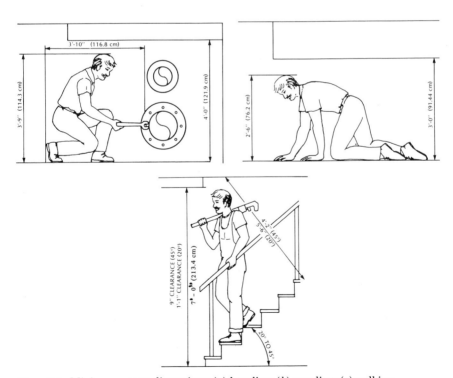

Figure 3.4 Minimum access dimensions: (*a*) kneeling; (*b*) crawling; (*c*) walking downstairs. *(Courtesy of Engineering Planning and Management, Inc.)*

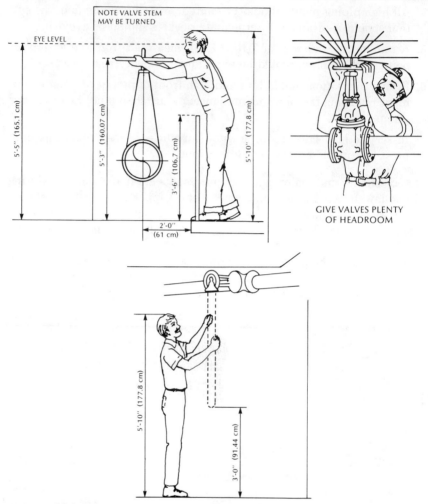

Figure 3.5 Maintenance pointers—convenient valve locations. *(Courtesy of Engineering Planning and Management, Inc.)*

lowing these recommendations permits maintenance crews to work within the comfort zones of their body heights. Additionally, valve operators and other equipment should avoid projecting into walkways where they might constitute safety hazards.

Flanged connections between piping and equipment should be provided to permit the dismantling which will occur during periodic maintenance, cleanup, and inspection. Piping should be arranged so as not to interfere with removal of equipment through hatches, manholes, doorways, or passageways.

3.4 Piping Design Drawings

Once the piping layout is determined, piping design drawings are prepared that show the routing. The piping drawing is the principal document used by the piping design groups. These drawings usually show both plan and elevation views. (A *plan view* is a view from above; an *elevation view* is a view seen from a point in the horizontal plane.)

A plan view of a piping system is shown in Fig. 3.6. This view shows the major vessels and piping as well as building penetrations. However, a plan view does not clearly show the changes in elevation; therefore, both plan and elevation views (Fig. 3.7) are needed to identify the full pipe routing. Piping systems are usually routed in straight sections combined with 90° elbows rather than following the shortest route to the point of termination. This permits the pipe route to follow the building structural steel and therefore simplifies support work.

Figure 3.7 illustrates an elevation view of the piping system. In this figure changes in the height of the piping system are more easily visualized. Some areas of the piping system that are not easily identified in the plan view can be clarified in the elevation view. For example, in Fig. 3.6, the elevation of the attachment to tank nozzle A is not easily identified. In Fig. 3.7, the change in elevation leading to the nozzle is more evident.

As seen in these figures, the piping is usually shown as a single solid

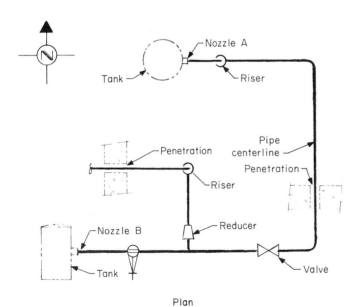

Figure 3.6 Plan view of piping.

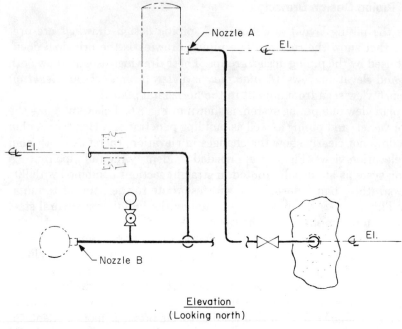

Figure 3.7 Elevation view of piping.

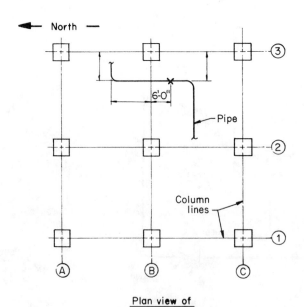

Figure 3.8 Typical piping reference axes for rectangular building (plan).

line. In some applications, larger-bore piping may be shown with the diameter drawn to scale in order to facilitate interference checks. The in-line components, such as valves, flanges, and reducers, are represented by symbols.

Once the pipe routing has been established, it must be located on the drawings by providing dimensions from easily accessible reference points. Piping is usually dimensioned from building structures such as walls or columns. Columns are usually identified as intersections of column lines, which are labeled by numbers in one direction and letters in the other, as shown in Fig. 3.8. In a circular building, points of interest either may be dimensioned from the center of the building or major equipment or may be identified by azimuth and radius, as shown in Fig. 3.9. When selecting a reference point for a dimension, one should consider that measurements taken will most probably begin at that location. Locations which are not easily accessible, such as the centerline of the storage pool, are not recommended reference points and should be avoided. If one location is dimensioned to column lines and all piping segment lengths are given on the drawing, then any other point on the pipeline can also be located.

Another reference for the piping system is the plant's north arrow. The north arrow is shown on the piping drawings and serves as a constant direction of reference for plant and piping layout. The north indicated is not necessarily true north; rather, it is a reference axis chosen specifically for the plant. The north arrow is usually selected to be parallel to one set of building column lines and thus to the structural steel. It is customary

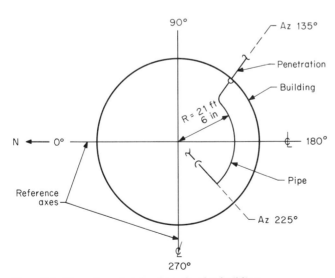

Figure 3.9 Plan view of piping in a circular building.

52 Piping and Pipe Support Systems

to choose the pipe routing either parallel or perpendicular to this north arrow where possible, because this usually ensures the optimum use of existing building structures for supports.

3.5 Composite Drawings and Scale Models

During the design phase of a project, there is a need for a composite layout coordinating the work of all disciplines. This is necessary to allocate space for all major equipment and to prevent interdisciplinary interferences. The various design groups use the composite drawings to lay out their portions of the work, thus ensuring the ability of each discipline to progress independently of any other.

Composite drawings are "layerings" of piping, equipment, and structural drawings found in a single area and include piping systems, cable

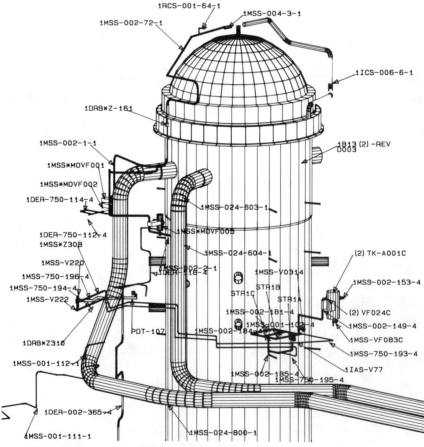

Figure 3.10 Piping and equipment composite. *(Courtesy of Construction Systems Associates, Inc., Marietta, Ga.)*

trays, HVAC systems, and other equipment. These drawings serve as design tools to ensure a coordinated and effective use of available space. A sample composite drawing is shown in Fig. 3.10. Composite drawings should be updated on a regular basis as design work continues.

Projects do not always use composite drawings as a design tool. Instead, a scale model may be used. The scale model is a reduced version of the plant under construction, including structures, major equipment, and piping. This visual representation assists in the design, fabrication, and installation of piping and supports.

The cost of a scale model may be as high as 0.1 percent of the total project cost. For example, if the cost of an 1100-MW nuclear plant is $3 billion, the scale model can cost as much as $3,000,000. This cost is expected to be offset by reduced design time and minimization of construction interference.

The model is similar to drawings in that it must show the most current piping and equipment locations. If design changes are drastic, the model may have to be reconstructed. Although the model is an excellent tool, it is not accurate enough for most design applications. Once designs are conceptualized by using the model, exact dimensions must be obtained from drawings.

3.6 Piping Isometric Drawings

Whereas the project design groups rely primarily on piping drawings as their sources, some project groups find it necessary to work from piping isometric drawings. Piping isometrics are three-dimensional representations of the piping shown in two dimensions on piping drawings. The piping isometric is used where conceptual layout is more important than exact scaled dimensions. Isometric drawings are commonly used for piping erection purposes and as stress analysis models.

Figure 3.11 shows sample isometric views of piping systems. The three-dimensional effect is created by representing the two horizontal axes (x and z) of the piping system 30° clockwise and 30° counterclockwise, respectively, from the horizontal axis of the paper while the vertical (y) axis conforms to the vertical axis of the paper. Piping that is not running parallel to *one of* the three major axes can be represented by showing its components along the other axes. From the figure, it is obvious that isometrics need not be drawn to scale; the piping segments may be drawn as long as is necessary for clarity.

The isometric drawing gives dimensions relative to the center of the pipe. The pipe centerline elevation is given at one point on the pipe. A vertical reference dimension is needed each time the pipe elevation changes. Dimensions along the pipe length are given in components parallel to the major axes of the building. The piping is dimensioned from column lines in order to specify its location in the building.

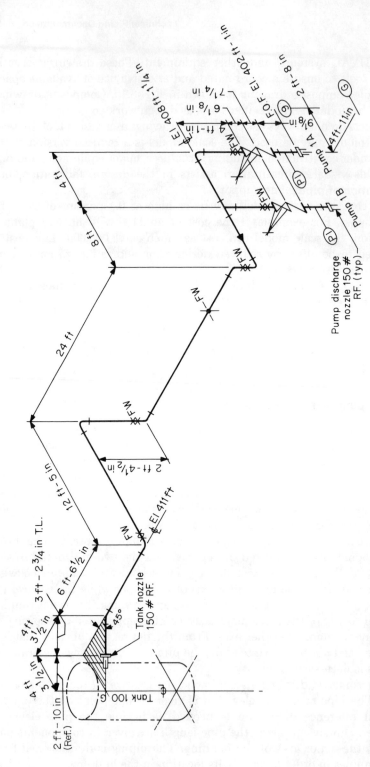

Figure 3.11 Typical piping isometric drawing.

An isometric shows the complete pipeline from one piece of equipment to another. It is usually prepared to facilitate pipe fabrication and assembly. In its completed form, the isometric may contain information concerning pipe fabrication and erection as well as the most current pipe support data. When they are used by the design, analysis, fabrication, and construction groups, isometrics give a clearer presentation of piping configuration than is available from plans and elevations.

Pipe supports are required to protect the piping system from the design loading conditions. Pipe design conditions can vary from the simple weight loads up to complex transient loads associated with earthquakes and pipe rupture. Initial selections of support locations and types are made based on design loadings, pipe size, system configuration, and building steel locations, with the design goal being the minimization of the number of supports and restraints. The initial support locations and types should be marked on the isometric drawing for use by the pipe stress analyst. The various types of supports as well as criteria for locating them are discussed in Chaps. 5, 6, and 7.

If, during the course of the stress analysis, fabrication, or installation, a support function, location, or direction of action changes, the isometric drawing should be updated to reflect this change.

3.7 Stress Isometrics

Stress isometrics are developed by using the piping drawings and isometrics as references. These isometric drawings serve as models for the stress analysis work and so must show all relevant information. As seen in the stress isometric in Fig. 3.12, the following items are expected to be included:

1. The global coordinate system should be shown with the positive linear and angular direction indicated for the x, y, and z reference axis.

2. The piping system must be located with respect to a reference building location. The piping centerline elevation and two horizontal dimensions from column lines are usually used in rectangular buildings while the elevation, radius, and azimuth are used in circular buildings.

3. Pipe node points should be selected at locations of interest, such as points of estimated high stress or deflection. The high-stress or deflection points may be estimated by reviewing the loading on the pipe span and pipe boundary conditions. Additional piping node points are located at concentrated loads (i.e. valves, etc.), support locations, design parameter transition points, and geometry intersection points. When a dynamic analysis is being done, it may be necessary to add node points to serve as mass points in the dynamic model. Each piping node must be numbered uniquely.

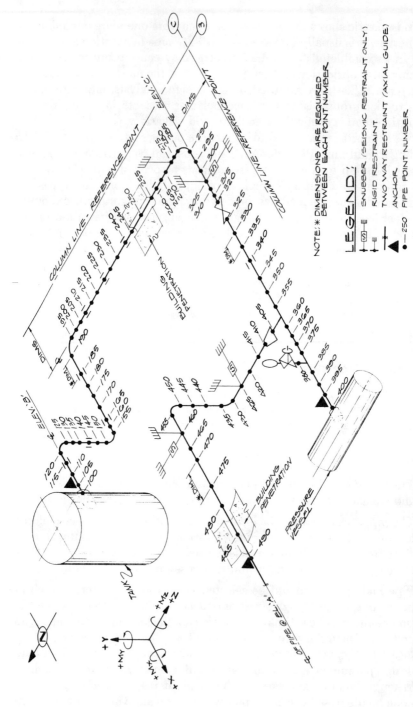

Figure 3.12 Sample pipe stress isometric. *(Courtesy of Impell Corporation.)*

4. The locations, functions, and lines of action of the supports should be represented.
5. Dimensions between node points should be shown, resolved into components parallel to each of the three global axes.
6. Other piping design parameters (such as pipe size, weight, temperature, pressure, material, valve weights, support stiffnesses, seismic acceleration factors, etc.) may be shown either on the isometric itself or on accompanying documents.

As noted previously, the stress analyst uses the piping isometric as a reference in constructing the stress isometric. Preliminary supports are located on the piping isometric; however, the stress analyst is in no way bound to use them all. It is preferable to use the minimum number of supports that permit all stress criteria to be met. The locations shown on the piping isometrics represent only those suggested locations where needed supports may be easily installed.

Supports (or restraints) are usually classified according to both direction and function. The major directions of restraint are those conforming to the three local axes of the pipe, although skewed restraints with components in more than one of these directions are occasionally found. The local axes for a pipe running in the horizontal plane are known as

- *Vertical:* Gravity direction
- *Axial:* Parallel to pipe run (sometimes called *longitudinal axis*)
- *Lateral:* Perpendicular to both the vertical and the axial axes (sometimes called *transverse axis*)

The function indicates the degree of restraint in terms of resistance to various loading cases. For example, a spring provides a force to counter a weight load, while providing little resistance to either thermal or dynamic loads. A snubber provides resistance to only dynamic loads, not to weight or thermal, while a rigid restraint resists all loads.

Colloquially, function names are also given to reflect the number of degrees of freedom restrained. For example, supports which are capable of resisting only downward loads are frequently called *hangers,* or *supports.* Supports that resist movement in all 6 degrees of freedom (three translational and three rotational) are normally called *anchors.* Anything in between, from 1 to 5 degrees of freedom, are known as *restraints.* However, this book uses the generic term *support* for all supports and restraints (for simplicity), except in those cases where a more specific term is required.

Supports may be built of varying types of hardware; however, their representation on the stress isometric is independent of hardware type. Conventions are sometimes developed that an anchor may be represented by a triangle (see Fig. 3.12, NP 100, 400, and 490), an X, or some other simple symbol.

Anchors are usually required at the beginning and end of stress analysis problems. The anchor is used as a location of transition between the pipe and equipment, the building structure, or simply the continuation of the pipe run. Forces and moments from one side of the anchor are not transmitted through to the piping beyond. Anchors may be added to stress analysis problems that are too large to handle in order to break up the large problem into two or more manageable problems. Such a use of anchors should be avoided, where possible, to reduce construction costs.

The pipe stress analyst uses the completed stress isometric as the model for the analysis performed, as discussed in Chap. 4. As significant changes occur to the piping and support configurations during further design or construction phases, the isometric will require revision, and it may be necessary to perform the stress analysis for the new conditions.

For nuclear safety-related piping and equipment, generally two types of reports are required when stress analyses are prepared. One is called a *seismic report* and is required for all quality assurance seismic category 1 components. This report confirms by either analysis or test data that the component can satisfy the code requirements applicable to seismic design. These requirements are defined in the project design specification. The second report is called a *stress report* and is required for ASME code, Section III, Class 1, Class MC, and Class 2 components designed according to NC-3200.

The stress report is a complete set of stress analysis calculations which establishes that the component satisfies all safety requirements and is adequate for all operating conditions, as specified by the rules of ASME code, Section III. Stress reports and seismic reports usually contain a checklist to ensure that compliance with all requirements has been reviewed and approved by the cognizant engineers. In the United States, stress reports must be approved by a professional engineer licensed in the state in which the nuclear plant is being constructed.

3.8 In-Service Inspection Drawings

Once operating, systems normally require in-service inspection to give early detection of potential failure, with the degree of attention given proportional to the criticalness of the system. In some systems, in-service inspection (ISI) is not performed until and unless leaks occur. However, for power piping, periodic inspection for maintenance purposes is prudent. For nuclear power plants, Section XI of the ASME Boiler and Pressure Vessel Code outlines the rules for in-service inspection. Piping design must reflect minimum clearances necessary for equipment that performs pressure boundary inspections.

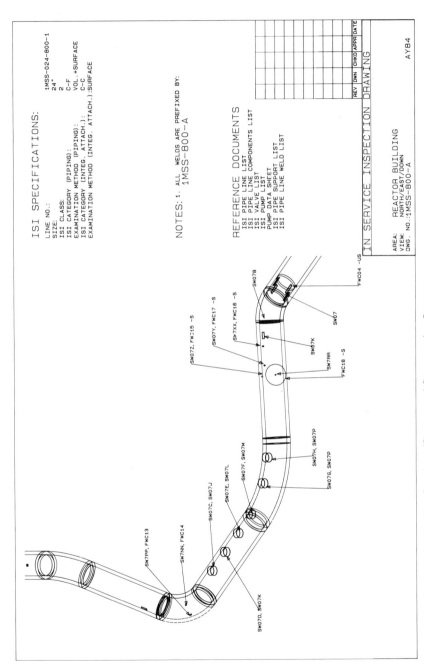

Figure 3.13 In-service inspection drawing. *(Courtesy of Construction Systems Associates, Inc., Marietta, Ga.)*

Drawings are prepared to assist the inspectors in their task, depicting welds and components to be examined. These will probably be the last piping documents to be finalized. Figure 3.13 illustrates a state-of-the-art computer-generated ISI drawing prepared for a main steam line. Locations along the pipe length where inspection is required (such as valves, fittings, instruments, lugs, etc.) are identified on this drawing.

Chapter 4

Overview of Pipe Stress Requirements

4.1 Introduction

In order to properly restrain a piping system the engineer should understand the requirements that are imposed on piping by the applicable codes, standards, and regulations. These requirements form the technical basis of the piping system design and analysis and therefore as a group comprise the design criteria. Chapter 2 introduced the codes, standards, and regulations enacted to protect the general public. Chapter 8 summarizes the specific code equations applicable to the design of the support itself. In this chapter, the design requirements embodied in some of the applicable pipe stress codes are discussed in more detail. If the engineer is not involved in the actual pipe stress analysis, this chapter need not be studied in detail.

The intent of the codes is to make known specific areas of concern and to identify requirements in the form of design principles. The ANSI Pressure Piping Code series and the ASME Boiler and Pressure Vessel Code, Section III, are the governing standards for the design of piping systems. They contain information the designer needs, such as allowable material stress values and design equations governing stresses resulting from pipe weight, temperature effects, and other design environments.

A majority of this chapter will deal with this information, extracted or summarized from the codes. However, it should not be construed that the procedures presented herein are universally accepted and applicable. The code information summarized here should not be applied without review by competent engineering personnel to determine its applicability to specific situations.

4.2 Theoretical Basis for the Piping Codes

4.2.1 Failure theories

U.S. piping codes usually subscribe to one of two failure theories in describing the strength of piping systems. The ANSI B31 codes as well as ASME Section III, Subsections NC and ND (Classes 2 and 3), have as their basis the *maximum principal stress theory,* making this theory the basis for most piping applications. This theory states that yielding occurs when the magnitude of any of the three mutually perpendicular principal stresses exceeds the yield strength of the material. This method has the advantage of being easy to apply; and when it is used with a suitable factor of safety, it yields acceptably safe results.

A more detailed evaluation of the state of stresses in a material coupled with a more accurate failure theory would permit the use of higher allowable stresses without a reduction in safety. This more accurate failure theory, which forms the basis for Subsection NB (Class 1) of ASME, Section III, is the *maximum shear stress theory.* This theory states that failure occurs when the maximum shear stress in a material exceeds the shear occurring in a uniaxial test sample at yield, or $S_y/2$.

The maximum shear stress τ_{max} is equal to one-half of the difference between the algebraically largest and smallest of the three principal stresses, as illustrated in the Mohr's circle shown in Fig. 4.1. In Fig. 4.1, σ_1, σ_2, and σ_3 are the principal stresses in the material, and τ_1, τ_2, and τ_3 are the shear stresses. Maximum shear stress $\tau_3 = (\sigma_1 - \sigma_3)/2$. So yielding occurs when

$$\tau_{max} = \frac{\sigma_{max} - \sigma_{min}}{2} = \frac{S_y}{2} \tag{4.1}$$

To simplify the mathematical operations required (by deleting the parallel divisions by 2), the Class 1 code defines a stress called the *equivalent intensity of combined stresses,* or *stress intensity.* This stress intensity S is defined as twice the maximum shear stress, and yielding is said to occur when the stress intensity exceeds the yield strength of the material.

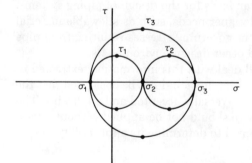

Figure 4.1 Mohr's circle, showing maximum shear stress.

Although both theories seem adequate for piping system design, the maximum shear stress theory seems more conservative than the maximum principal stress theory since it considers the contributory effects of orthogonal stresses. However, if all three principal stresses in a material were equal, the shear stresses calculated would be zero, and the maximum shear stress theory would allow unlimited stresses. For this reason, the Class 1 piping code also provides a limit on principal stresses.

4.2.2 Stress categories

Various failure modes could affect a piping system, including buckling, stress corrosion, and brittle fracture. These topics are not currently covered by the piping codes, but the effects of these failure modes must be considered by the piping engineer when selecting materials or laying out and restraining the piping system.

The failure modes that the piping codes address are excessive plastic deformation or bursting; plastic instability, or incremental collapse due to cycling in the plastic range; and high-strain low-cycle fatigue. Each of these failure modes is caused by a different type of stress and loading, so it has been necessary to separate them into three different categories and set different limits for each. The piping codes have broken the types of stresses, and the failure modes they guard against, into the following categories:

- *Primary stress.* Plastic deformation and bursting
- *Secondary stress.* Plastic instability leading to incremental collapse
- *Peak stress.* Fatigue failure collapse resulting from cyclic loadings

Primary stress is developed by imposed mechanical loadings (forces). Primary stress is not self-limiting. Therefore, if the yield strength is exceeded through the entire cross section of the structural material used in piping design, then failure can be prevented only by removal of the loading or strain hardening in the material.

Primary stresses may be further categorized as general primary membrane stress, local primary membrane stress, and primary bending stress. These three categories are of interest because the pipe will not fail until the entire cross section has reached the yield strength. Local primary stresses may exceed yielding; however, under this stress state they will behave as secondary stresses and redistribute themselves as the local pipe wall distortion occurs. The failing moment would be that required to put the entire cross section of the pipe in plastic behavior, not just the extreme fiber. Therefore, the permissible primary bending moment (and likewise the calculated stress) may be increased over the yielding moment by the shape factor.

64 Piping and Pipe Support Systems

Secondary stress is developed in a structure owing to constraint of that structure against displacements, whether thermal expansion or imposed anchor and restraint movements. Under secondary loading the piping system must conform to imposed strains, rather than imposed forces, so that the loading can be satisfied by system distortions. Distortion of the piping system as well as local yielding tends to relieve the developed stresses due to imposed displacements, so these stresses are said to be self-limiting.

Given that secondary stresses are classified as those stresses caused by constraints of displacements which cause distortion, peak stresses are those which cause virtually no distortion and therefore high stress levels. Examples would be thermal gradients through a pipe wall or stress concentration at a discontinuity such as a pipe fitting or a weld. *Peak stress is the highest stress in a local region and is responsible for causing fatigue failure.*

4.2.3 Stress limits

The limits for each stress type have been determined by the application of limit theory (assuming perfectly elastic and plastic material behavior) coupled with suitable factors of safety.

As mentioned previously, a pipe is assumed to fail when an applied loading induces a general primary membrane stress equal to the yield

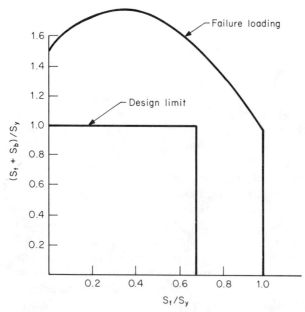

Figure 4.2 Failure loading versus design limits (rectangular section).

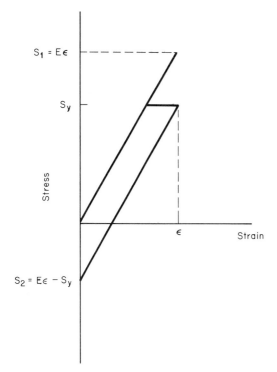

Figure 4.3 Stress-versus-strain diagram.

strength of the material. Failure under a moment loading requires that the total cross section be at the yield strength; this will not occur until the moment is increased above the yielding moment of the pipe multiplied by the shape factor. Combination of these primary membrane and bending loadings would result in reduced allowables for both bending and membrane stress, as shown in Fig. 4.2, in which S_t = tensile stress, S_b = bending stress, and S_y = yield stress. The stress limits usually defined for general primary membrane stress ($\frac{2}{3}S_y$) and for combined calculated primary stress (S_y) have been shown to provide an adequate safety margin to prevent yielding failure.

Limits for secondary stresses are given in terms of allowable calculated stress range. The secondary stresses, when combined with primary stresses or even when taken alone, are permitted to occasionally stress the pipe beyond its yield point. The reason for this allowable overstress is that a repetitively applied load which initially forces the pipe into the plastic range will, after a few cycles, "shake down" and be reduced to elastic action in the piping system.

This stress theory can be understood by considering a pipe experiencing an imposed displacement which is beyond its yield strain. Figure 4.3, in which S_1 = calculated stress and S_2 = residual stress due to plastic deformation, shows the calculated stress value would be equal to the product of the modulus of elasticity E and the strain ϵ, resulting in a value which exceeds the yield strength. When the cyclic displacement is removed, the pipe location that exceeded the yield point will retain a residual distortion equal to the imposed strain less the yield strain. This distortion will induce a residual stress, opposite in nature to that developed during loading and equal to the difference between the calculated stress value and the material yield stress. This residual stress, tension stress, for example, must be overcome on each subsequent loading cycle before the pipe can go into compression. The elastic—and therefore, the allowable—range has been extended by the value of the prestress, or $E\epsilon - S_y$. For this reason, the allowable secondary stress range may be as high as $2S_y$ (times a suitable factor of safety).

This piping "shakedown" is also known as *self-springing,* and it can be hastened by the process known as *cold-springing,* in which a pipe is intentionally cut and installed short to provide the prestress normally provided by plastic deformation. This pretensioning of the piping system has the effect of reducing the calculated thermal stresses (and the actual thermal restraint loads) in the system; however, there is no difference in the stress *range* to the pipe, whether it is cold-sprung or self-sprung. The calculated secondary stress range in a system may not be reduced through the use of any kind of prestress.

4.2.4 Fatigue

As mentioned previously, the third type of stress, *peak stress,* is caused by localized discontinuities, is a secondary stress, and produces virtually no distortion. For these reasons, peak stresses cannot cause failure from bursting or yielding. However, the continued cycling of high-stress concentrations may eventually cause a crack which propagates and results in leakage. This failure mechanism is called *fatigue.*

Note that the term *fatigue,* which was adopted well over a century ago, may not be the best choice of word. The actual aspects of this phenomenon are different from the process of biological fatigue. First, fatigue failure may occur with little or no warning. Also, periods of rest with the fatigue stress removed will not lead to any healing or recovery from the effects of the prior stress. Thus, damage once done during the fatigue process is cumulative and normally unrecoverable.

Fatigue failure is usually classified in textbooks under two different stress characteristics, and in both cases the failure is caused by different physical mechanisms. With one type of fatigue, significant plastic strain occurs during each cycle. This type of fatigue, called *low-cycle fatigue,* is characterized by high loads and a small number of cycles before failure.

The second type of fatigue results from strain cycles in the elastic range; this mechanism is called *high-cycle fatigue*. Low-cycle fatigue typically includes those failures occurring within 10^5 cycles or fewer.

Fatigue failure has long been understood and used in the design of rotating machinery; this type of fatigue is of the high-cycle type. In high-cycle fatigue analysis, a stress level known as the *endurance limit*, which may be applied an infinite number of times without failure, is calculated. In piping design, most of the loading cycles encountered would be of the low-cycle type, being expected to occur less than 10^5 times during the service life of the system. For low-cycle fatigue, failure will occur only with stress levels in the plastic range. The stresses which cause fatigue failure in the piping are the peak stresses.

For every material, a fatigue curve can be generated by experimental analysis which correlates peak stress range with the number of cycles to failure. A typical fatigue curve, also known as the *S-N curve*, is shown in Fig. 4.4. The alternating stress S_A is defined as one-half of the calculated peak stress. By ensuring that the number of load cycles N that the system experiences is fewer than the number permitted for the alternating stress developed, fatigue failure may be prevented.

In piping applications, the alternating stress may be expected to vary in magnitude during the system service life. These variations make the direct use of the fatigue curves inapplicable since the curves are developed for constant-stress amplitude operation. Therefore, it is important for the engineer to have a theory or hypothesis, verified by experimental observations, that will permit use of the curve for the application.

One hypothesis asserts that the damage fraction of any stress level S_i is linearly proportional to the ratio of the number of cycles of operation

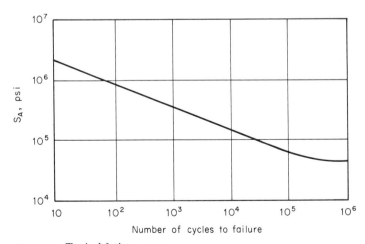

Figure 4.4 Typical fatigue curve.

at the stress level to the total number of cycles that would produce failure at that stress level. This means that failure is predicted to occur if

$$U = 1.0 \tag{4.2}$$

where $U =$ usage factor $= \sum \dfrac{n_i}{N_i}$ for all stress levels

n_i = number of cycles operating at stress level i

N_i = number of cycles to failure at stress level i as per material fatigue curve

This theory is easy to use and is acceptable, since more complex cumulative-damage theories do not always yield a significant improvement in failure prediction reliability.

4.3 Occasional Loads and Service Levels

The piping codes separate primary loads and stresses into two types based on the duration of their application. The first type is called *sustained loads,* or loads which can be expected to be present virtually at all times of plant operation. Examples of sustained loads would be weight and pressure loadings associated with the piping system's normal operating conditions. The second type is called *occasional loads,* or loads which are present during only a small fraction of the piping system operating time. Examples of occasional loads are high winds, fluid hammer, relief valve discharge, earthquake, and high-energy pipe break.

Service levels are defined for nuclear safety-related piping by the ASME code, Section III. They are, in order of decreasing likelihood and increasing severity of occurrence, levels A, B, C, and D, also known as *normal, upset, emergency,* and *faulted,* respectively. The piping codes regulate the stress levels permitted, but they do not define types of loading to be considered under each service level. The responsibility for determining the loadings to be considered under each service level rests with the plant owner upon advice of the architect or engineer and concurrence of the NRC. The definition of loading combinations would be recorded in the project specification or, in the case of a nuclear plant, in the safety analysis report.

Level A loadings refer to design conditions to which the piping system may be subjected during the performance of its specified service function. These would normally be loads due to operating pressure and weight.

Level B loadings include those occasional loadings which the piping system must withstand without suffering damage requiring repair. Usual examples of level B loadings would include fluid hammer, relief valve discharge, and *operating basis earthquake* (OBE), defined as the maximum earthquake postulated to occur within the design lifetime of the plant or one-half of the safe shutdown earthquake (see level C), whichever is greater. The codes allow increased stress levels for level B loadings, but not sufficient to allow damage to occur.

Level C loadings are normally those loadings associated with the design accidents of the plant. During level C loadings, the systems must be capable of performing their safety functions to safely shut down the plant. It is not expected that any damage will occur during these loading events; however, the piping components would be subject to inspection and repair, if necessary, prior to resumption of service after a level C loading. A common level C loading is *safe shutdown earthquake* (SSE), which is defined as the maximum earthquake postulated to occur at the plant site at any time. The limits set by the codes for level C loadings are high enough to permit large deformation in areas of structural discontinuity. Therefore, dynamic damping values are usually increased for this design condition.

Level D loadings are those associated with the most extreme accidents and less probable design conditions, such as a loss of coolant accident. For this loading level, yet higher loads and resulting damage and deformations are postulated in the piping. As long as the components retain their ability to perform their safety function, an increase in piping stress is usually permitted.

4.4 B31 Committee Projects

As described in Chap. 2, the need for a national code for pressure piping became evident from 1900 to 1925. To meet this need, project B31 was initiated in March 1926, at the request of the ASME, with that society as sole sponsor. After several years of issuing a single code, the committee was split into separate industry committees (i.e., power piping, fuel gas piping, etc.), with each issuing its own code. The ASME has adopted these codes, and they are published jointly as ANSI/ASME codes.

The codes contain basic reference data and formulas necessary for design, including

1. Material specifications and component standards which are acceptable for code usage
2. Acceptable dimensional standards for the elements comprising piping systems
3. Requirements for the pressure design of component parts and assembled units
4. Requirements for the evaluation and limitation of stresses, reactions, and movements associated with pressure, temperature, and external forces

5. Requirements for the fabrication, assembly, and erection of piping systems
6. Requirements for examination, inspection, and testing of piping systems

These requirements may not be followed selectively, but must be adhered to as a whole. The following sections of this chapter provide summaries of the stress requirements contained in the more complex codes. Where metric units are not in the original code, conversion has been by the authors. These summaries are for instructional purposes only, and the reader is cautioned to obtain a copy of the complete code of interest, prior to application of any code requirements.

4.4.1 B31.1 Power Piping Code

The scope of the B31.1 Power Piping Code prescribes minimum requirements for the design, materials, fabrication, assembly, erection, test, and inspection requirements for power and auxiliary service piping systems for electric generating stations, industrial and institutional plants, central and district heating plants, and district heating systems both on the property of and within buildings of the users. Power piping services include, but are not limited to, steam, water, oil, gas, and air services.

Design requirements discussed in this code include those for pipe, flanges, gaskets, valves, relief devices, fittings, and the pressure-containing parts of other piping components. It also covers restraints and other equipment necessary to prevent overstressing the pressure-containing components.

The loadings required to be considered for B31.1 are pressure, weight (live, dead, and under test loads), impact (fluid hammer, for example), wind, earthquake (where applicable), vibration, and those loadings resulting from thermal expansion and contraction. Earthquake and wind loading are not usually considered to occur simultaneously in this code.

The equations to be satisfied in order to meet the requirements of the ANSI B31.1 code are extracted directly from the code as follows. Note that both U.S. Customary System (USCS) units and metric, or International System (abbreviated SI), units are given.

Stresses due to sustained loadings. The effects of pressure, weight, and other sustained mechanical loads must meet the requirements of the following equation:

$$\frac{PD_o}{4t_n} + \frac{0.75iM_A}{Z} \leq 1.0S_h \quad \text{(USCS units)}$$

$$\frac{PD_o}{4t_n} + \frac{1000(0.75i)M_A}{Z} \leq 1.0S_h \quad \text{(SI Units)}$$

(4.3)

where P = internal design pressure (gauge), psi (kPa)
D_o = outside diameter of pipe, in (mm)
t_n = nominal wall thickness of pipe, in (mm)
M_A = resultant moment loading on cross section due to weight and other sustained loads, in·lb (mm·N)
Z = section modulus of pipe, in^3 (mm^3)
i = stress intensification factor (the product $0.75i$ shall never be taken as less than 1.0)
S_h = basic material allowable stress at maximum temperature from allowable stress tables, psi (kPa)

Stresses due to occasional loadings. The effects of pressure, weight, other sustained loads, and occasional loads including earthquake must meet the following requirements:

$$\frac{PD_o}{4t_n} + \frac{0.75iM_A}{Z} + \frac{0.75iM_B}{Z} \leq kS_h \quad \text{(USCS)}$$

$$\frac{PD_o}{4t_n} + \frac{1000(0.75i)M_A}{Z} + \frac{1000(0.75i)M_B}{Z} \leq kS_h \quad \text{(SI)}$$

(4.4)

where k = 1.15 for occasional loads acting less than 10 percent of operating period
= 1.2 for occasional loads acting less than 1 percent of operating period
M_B = resultant moment loading on cross section due to occasional loads; if earthquake is required, only one-half the earthquake moment range should be used, in·lb (mm·N)

Stress range due to expansion loadings. The effects of thermal expansion stresses must satisfy the following equation:

$$S_E = \frac{iM_c}{Z} \leq S_A + f(S_h - S_L) \quad \text{(USCS)}$$

$$S_E = \frac{1000iM_c}{Z} \leq S_A + f(S_h - S_L) \text{ (SI)}$$

(4.5)

where M_c = range of resultant moments due to thermal expansion, in·lb (mm·N)
S_A = allowable stress range for thermal expansion stresses
= $f(1.25S_c + 0.25S_h)$, psi (kPa)
S_c = basic material allowable stress at minimum (cold) temperature from the allowable stress tables, psi (kPa)
f = stress range reduction factor for cyclic conditions during operating modes (see Table 4.1)
S_L = sustained stress calculated in Eq. (4.3)

TABLE 4.1 Stress Range Reduction Factor f

Number of equivalent temperature cycles	f
<7000	1.0
7000– 14,000	0.9
14,000– 22,000	0.8
22,000– 45,000	0.7
45,000–100,000	0.6
>1000,000	0.5

SOURCE: B31.1 Piping Code. *(Courtesy of ASME.)*

This code requires that the piping be designed for a metal temperature representing the maximum temperature expected. The B31.1 code states that the design temperature shall be assumed to be the same as the fluid temperature unless calculations are provided to support other values. Under no circumstances shall the design temperatures be less than the average of the fluid temperature and the outside pipe wall temperature.

The basic material allowable stresses to be used for S_c and S_h for most materials are determined by selecting the minimum value of the following possible design stresses: (1) 0.25 times the ultimate tensile strength of materials used at design temperature, (2) 0.25 times ultimate tensile strength at installation temperature, (3) 0.625 times the yield strength at design temperature, or (4) 0.625 times the yield strength at installation temperature. These values are tabulated for a wide range of materials and operating temperatures in Appendix A of the B31.1 code.

The stress intensification factor i is applied as an additional factor of safety at fittings, welds, and other locations where stress concentrations and possible fatigue failure could occur. The B31.1 code requires the use of a common intensification value for both in-plane and out-of-plane moments as well as for torsion. The formulas for stress intensification factors are given in Appendix D of the code and are also shown in Table 4.2.

4.4.2 B31.3 Chemical Plant and Petroleum Refinery Piping Code

The scope of Code B31.3 Chemical Plant and Petroleum Refinery Piping encompasses all piping within the property limits of facilities processing or handling chemical, petroleum, or related products. A few examples are chemical plants, petroleum refineries, loading terminals, and natural gas processing plants. Excluded from B31.3 are piping carrying nonhazardous fluids with an internal gauge pressure lower than 15 psi (103 kPa) and a temperature below 366°F (186°C); steam and feedwater piping per ANSI B31.1, B31.4, or B31.8; and plumbing, sewers, and fire protection systems.

The loadings required to be considered for B31.3 are pressure, weight (live and dead loads), impact (including hydraulic shock), wind, earth-

TABLE 4.2 Stress Intensification Factors

Component description	As per B31.1, ASME, III NC and ND	As per B31.3 and B31.8 Out-of-plane i_o	As per B31.3 and B31.8 In-plane i_i	Flexibility characteristic h	Sketch
Welding elbow or pipe bend	$\dfrac{0.9}{h^{2/3}}$	$\dfrac{0.75}{h^{2/3}}$	$\dfrac{0.9}{h^{2/3}}$	$\dfrac{\overline{T} R_1}{r_2^2}$	
Closely spaced miter bend $s < r_2(1+\tan\theta)$ $(s = 2r_2 \tan\theta)$	$\dfrac{0.9}{h^{2/3}}$	$\dfrac{0.9}{h^{2/3}}$	$\dfrac{0.9}{h^{2/3}}$	$\dfrac{\cot\theta}{2}\dfrac{\overline{T} s}{r_2^2}$	
Single miter bend or widely spaced miter bend $s \geq r_2(1+\tan\theta)$	$\dfrac{0.9}{h^{2/3}}$	$\dfrac{0.9}{h^{2/3}}$	$\dfrac{0.9}{h^{2/3}}$	$\dfrac{1 + \cot\theta\,\dfrac{\overline{T}}{r_2}}{2}$	
Welding tee per ANSI B16.9	$\dfrac{0.9}{h^{2/3}}$	$\dfrac{0.9}{h^{2/3}}$	$0.75 i_o + 0.25$	$\dfrac{4.4\overline{T}}{r_2}$	
Reinforced fabricated tee with pad or saddle	$\dfrac{0.9}{h^{2/3}}$	$\dfrac{0.9}{h^{2/3}}$	$0.75 i_o + 0.25$	$\dfrac{(\overline{T} + \tfrac{1}{2} t_r)^{5/2}}{\overline{T}^{3/2} r_2}$	

Unreinforced fabricated tee	$\dfrac{0.9}{h^{2/3}}$	$\dfrac{0.9}{h^{2/3}}$	$0.75i_o + 0.25$	$\dfrac{\overline{T}}{r_2}$
Extruded welding tee	Not given	$\dfrac{0.9}{h^{2/3}}$	$0.75i_o + 0.25$	$\left(1 + \dfrac{r_x}{r_2}\right)\dfrac{\overline{T}}{r_2}$
Welded in contour insert	Not given	$\dfrac{0.9}{h^{2/3}}$	$0.75i_o + 0.25$	$\dfrac{4.4\overline{T}}{r_2}$
Branch welded on fitting (integrally reinforced)	Not given	$\dfrac{0.9}{h^{2/3}}$	$0.75i_o + 0.25$	$\dfrac{3.3\overline{T}}{r_2}$

SOURCE: Courtesy of ASME.

quake-induced horizontal forces, vibration fluid discharge reactions, thermal loadings due to expansion and contraction, thermal gradients, and loadings due to support, anchor, and terminal movements. The requirements of the ANSI B31.3 code are as follows:

Stresses due to sustained loadings. The sum of the longitudinal stresses S_L due to pressure, weight, and other sustained loadings shall not exceed S_h. When S_L is calculated, the thickness of the pipe used shall not include allowances for corrosion, erosion, threads, or groove depth, where S_h = basic allowable stress at maximum metal temperature.

Stresses due to occasional loadings. The sum of the longitudinal stresses due to pressure, weight, and other sustained loadings and of the stresses produced by occasional loads such as wind or earthquake must not exceed $1.33S_h$. Wind and earthquake forces need not be considered as acting concurrently.

Stress range due to expansion loadings. The displacement stress range S_E shall be computed according to

$$S_E = (S_b^2 + 4S_t^2)^{1/2} \tag{4.6}$$

where S_b = resultant bending stress, psi (kPa)

$$S_t = \frac{M_t}{2Z} = \text{torsional stress, psi}$$

$$= \frac{1000 M_t}{2Z} \text{ kPa}$$

M_t = torsional moment, in·lb (mm·N)
Z = section modulus of pipe, in³ (mm³)

The bending stresses S_b to be used in Eq. (4.6) shall be calculated as

$$S_b = \left[\frac{(i_i M_i)^2 + (i_o M_o)^2}{Z}\right]^{1/2} \quad \text{(USCS)} \tag{4.7}$$

$$S_b = 1000 \left[\frac{(i_i M_i)^2 + (i_o M_o)^2}{Z}\right]^{1/2} \quad \text{(SI)}$$

where i_i = in-plane stress intensification factor
i_o = out-of-plane stress intensification factor
M_i = in-plane bending moment, in·lb (mm·N)
M_o = out-of-plane bending moment, in·lb (mm·N)

The following displacement stress range limit must be maintained:

$$S_E \leq S_A \tag{4.8}$$

where $S_A = f(1.25 S_c + 0.25 S_h)$, psi (kPa)
or $\quad = f[1.25(S_c + S_h) - S_L]$, psi (kPa)

S_c = basic allowable stress at minimum metal temperature, psi (kPa)

S_L = calculated longitudinal stresses due to pressure, weight, and other sustained loadings, psi (kPa)

f = stress range reduction factor for cyclic conditons for total number of full-temperature cycles over expected life, per Table 4.1

The basic material allowable stresses to be used for S_c and S_h for most materials are determined by selecting the minimum value of: (1) one-third of the specified minimum tensile strength at room temperature, (2) one-third of the tensile strength at design temperature, (3) two-thirds of the specified minimum yield strength at room temperature, (4) two-thirds of the yield strength at design temperature (except for austenitic stainless steels and certain nickel alloys, where this value may be as high as $0.9 S_y$ at temperature), (5) 100 percent of the average stress for a 0.01 percent creep rate per 1000 h (6) 67 percent of the average stress for rupture at the end of 100,000 h, and (7) 80 percent of the minimum stress for rupture at the end of 100,000 h. These values have been tabulated for approved materials and operating temperatures in Appendix A of the code.

The B31.3 code requires the use of separate stress intensification factors for in-plane and out-of-plane moment loadings, with no intensification required for torsional moments. The formulas for the required intensification factors are given in Table 4.2.

4.4.3 B31.7 Nuclear Power Piping Code

ANSI B31.7 was the original Nuclear Power Piping Code which has since been superseded by Section III of the ASME Boiler and Pressure Vessel Code. Despite the withdrawal, this code still may see use during backfit work on older nuclear plants which had B31.7 as the code of record.

The scope of this code covered those piping systems designed to provide a pressure-retaining barrier for safety-related systems in nuclear power plants. Excluded from this code is piping specifically covered by other sections of the pressure piping code (such as the conventional steam piping portion of a nuclear plant, which would be designed according to B31.1); piping used specifically for processing nuclear fuels; some parts of piping for marine installations subject to supplementary requirements under marine regulatory agencies; building heating and distribution steam piping with a gauge pressure less than 15 psi (103 kPa) or hot-water heating systems with a gauge pressure less than 30 psi (206 kPa); nonnuclear roof and floor drains, plumbing, sewers, fire protection systems, and piping for hydraulic or pneumatic components of tools and equipment.

ANSI B31.7 introduced safety classes to nuclear piping, calling for separate criteria for Class 1, 2, and 3 piping (in descending order of criticalness). According to the code, the responsibility for establishing the appropriate classification for all nuclear piping fell to the owner. Since the specific requirements of ANSI B31.7 are similar to those of ASME Section III (and since the code is superseded), they are not discussed here.

4.4.4 B31.8 Gas Transmission and Distribution Piping Code

The scope of Code B31.8 Gas Transmission and Distribution Piping encompasses pipelines in gas distribution systems up to customers' meters. Excluded from this code is piping with metal temperatures below $-20°F$ ($-29°C$) or above $450°F$ ($232°C$); piping beyond the customers' meters; piping in oil refineries, natural gas extraction plants, gas treating plants, etc., that is covered by other ANSI B31 code sections; vent piping operating at atmospheric pressures for waste gases; and liquid-petroleum piping systems.

Stresses due to primary loadings. The sum of the longitudinal pressure stress and the longitudinal bending stress due to external loads such as weight of pipe and contents, wind, etc., shall be limited to

$$S_L \leq 0.75 SFT \tag{4.9}$$

where S = specified minimum yield strength, psi (kPa)

T = temperature derating factor, for steel (see Table 4.3)

F = construction type factor:

Construction type	Design factor F
A	.72
B	.60
C	.50
D	.40

These construction types are defined in Section 841 of the code. They are approximately dictated by the population density of the surrounding area: type A is found in sparsely populated areas (mountains, deserts, farmland, etc.), type B is found in fringe areas around cities or towns, type C in cities or towns with no buildings over three stories tall, and type D in areas with taller buildings.

TABLE 4.3 Temperature Derating Factor T

Temperature	T
250°F (121°C) or less	1.000
300°F (149°C)	0.967
350°F (177°C)	0.933
400°F (204°C)	0.900
450°F (232°C)	0.867

SOURCE: B31.8 Piping Code. *(Courtesy of ASME.)*

Stress range due to expansion loading. The expansion stress range S_E shall be calculated as and limited to

$$S_E = (S_b^2 + 4S_t^2)^{1/2} \leq 0.72S \tag{4.10}$$

where S_b = resultant bending stress = iM_b/Z, psi
 = $1000iM_b/Z$, kPa
S_t = torsional stress = $M_t/(2Z)$, psi
 = $500M_t/Z$, kPa
M_b = resultant bending moment, in·lb (mm·N)
M_t = torsional moment, in·lb (mm·N)
Z = section modulus of pipe, in^3 (mm^3)
i = stress intensification factor
S = specified minimum yield strength, psi (kPa)

Stresses due to primary plus expansion loadings. The sum of the expansion stress range, the longitudinal pressure stress, and the longitudinal bending stress due to primary loadings shall not exceed the specified minimum yield strength S.

ANSI B31.8 uses in-plane and out-of-plane stress intensification factors, for which formulas are given in Table 4.2.

4.4.5 ASME Boiler and Pressure Vessel Code, Section III, Subsection NB

Subsection NB of the ASME Boiler and Pressure Vessel Code, Section III, details the requirements pertaining to those sections of nuclear piping designated as Class 1. This piping code is an outgrowth of, and has superseded, the Class 1 section of the previously discussed B31.7 code.

The loadings requiring consideration in the design of piping under subsection NB are pressure, weight (live and dead loads), impact, earthquake, vibration, and loadings induced by thermal expansion and contraction. The stress requirements which must be met to satisfy the Class 1 code are as follows:

Primary stress intensity check. The primary stress intensity limit is satisfied when

$$\frac{B_1 P D_o}{2t} + \frac{B_2 D_o M_i}{2I} \leq kS_m \quad \text{(USCS)}$$
$$\frac{B_1 P D_o}{2t} + \frac{1000 B_2 D_o M_i}{2I} \leq kS_m \quad \text{(SI)} \tag{4.11}$$

where B_1, B_2 = primary stress indices for specific component under investigation

P = design gauge pressure, psi (kPa)
D_o = outside diameter of pipe, in (mm)
t = nominal wall thickness, in (mm)
I = moment of inertia, in^4 (mm^4)
M_i = resultant moment due to combination of design mechanical loads, in·lb (mm·N)
k = 1.5 for level A, 1.8 for level B, 2.25 for level C
S_m = allowable design stress intensity value, psi (kPa)

Primary plus secondary stress intensity range. This evaluates a stress range as the system goes from one load set (pressure, temperature, moment, and force loading) to any other load set which follows in time. It is the range of pressure, temperature, and moments between the two load sets which is to be used in the calculations. For each specified pair of load sets, calculate S_n:

$$S_n = \frac{C_1 P_o D_o}{2t} + \frac{C_2 D_o M_i}{2I} + C_3 E_{ab} |\alpha_a T_a - \alpha_b T_b| \qquad \text{USCS}$$

$$S_n = \frac{C_1 P_o D_o}{2t} + \frac{1000 C_2 D_o M_i}{2I} + C_3 E_{ab} |\alpha_a T_a - \alpha_b T_b| \qquad \text{SI}$$

(4.12)

where C_1, C_2, C_3 = secondary stress indices for component under investigation

M_i = resultant range of moment which occurs when system goes from one service load set to another, in·lb (mm·N)

T_a, T_b = range of average temperature on side a or b of gross structural discontinuity or material discontinuity, °F (°C)

$\alpha_{a,b}$ = coefficient of thermal expansion on side a or b of a gross structural discontinuity or material discontinuity at room temperature, in/(in·°F) [mm/(mm·°C)]

E_{ab} = average modulus of elasticity of two sides of gross structural discontinuity or material discontinuity at room temperature, psi (kPa)

P_o = range of service pressure, psi (kPa)

The following limit applies:

$$S_n \leq 3 S_m \qquad (4.13)$$

If this equation is not satisfied for all pairs of load sets, then the component may still be qualified by using the simplified elastic-plastic discontinuity analysis described below; otherwise, the analyst may proceed to the peak-stress range calculation.

Simplified elastic-plastic discontinuity analysis. If S_n exceeds its limit for some pairs of load sets, a simplified elastic-plastic analysis may be performed if thermal stress ratchet is not present. This analysis is required only for the specific load sets that exceeded the primary plus secondary stress intensity range check. The following two sets of equations must be satisfied:

$$S_e = \frac{C_2 D_o M_i^*}{2I} \leq 3S_m \quad \text{(USCS)}$$
$$S_e = \frac{1000 C_2 D_o M_i^*}{2I} M_i^* \leq 3S_m \quad \text{(SI)} \tag{4.14}$$

where S_e = expansion stress and M_i^* = resultant range of moments due to thermal expansion and thermal anchor movements, in·lb (mm·N), and

$$\frac{C_1 P_o D_o}{2t} + \frac{C_2 D_o M_i}{2I} + C_3' E_{ab} |\alpha_a T_a - \alpha_b T_b| \leq 3S_m \quad \text{(USCS)}$$
$$\frac{C_1 P_o D_o}{2t} + \frac{1000 C_2 D_o M_i}{2I} + C_3' E_{ab} |\alpha_a T_a - \alpha_b T_b| \leq 3S_m \quad \text{(SI)} \tag{4.15}$$

where M_i = resultant range of moment which occurs when system goes from one service load set to another, excluding moments due to thermal expansion and thermal anchor moments, in·lb (mm·N)

C_3' = stress index for component under investigation

If $S_n > 3S_m$, the thermal stress ratchet must be evaluated and demonstrated to be satisfactory before a simplified elastic-plastic discontinuity analysis can be performed. This ratchet is a function of the $|\Delta T_1|$ range only. The following equation must be satisfied:

$$|\Delta T_1| \text{ range} \leq \frac{y' S_y C_4}{0.7 E \alpha} \tag{4.16}$$

where y' = 3.33, 2.00, 1.20, and 0.80 for x = 0.3, 0.5, 0.7, and 0.8, respectively

$$x = \frac{P D_o}{2t} \frac{1}{S_y}$$

P = maximum pressure for conditions under consideration, psi (kPa)

C_4 = 1.1 for ferritic material and 1.3 for austenitic material

E = modulus of elasticity at room temperature, psi (kPa)

S_y = yield strength value taken at average fluid temperature under consideration psi (kPa)

α = coefficient of thermal expansion at room temperature, in/(in·°F) [mm/(mm·°C)]

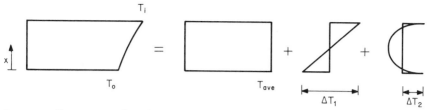

Figure 4.5 Temperature distribution in a pipe wall.

Peak stress intensity range and fatigue analysis. Please refer to Fig. 4.5, in which T_i = internal surface temperature, T_o = external surface temperature, T_{ave} = average temperature of pipe wall, $\Delta T_1 = T_i - T_o$, and $\Delta T_2 = (-1/t)\int_o^t T(x)\,dx + (T_o + T_i)/2$, where t = wall thickness and $T(x)$ = temperature as function of distance through wall. For each loading condition specified, peak stress values must be calculated:

$$S_p = \frac{K_1 C_1 P_o D_o}{2t} + \frac{K_2 C_2 D_o M_i}{2I} + \frac{1}{2(1-\nu)} K_2 E\alpha |\Delta T_1|$$

$$+ K_3 C_3 E_{ab}|\alpha_a T_a - \alpha_b T_b| + \frac{1}{1-\nu} E\alpha |\Delta T_2| \quad \text{(USCS)}$$

$$S_p = \frac{K_1 C_1 P_o D_o}{2t} + \frac{1000 K_2 C_2 D_o M_i}{2I} + \frac{1}{2(1-\nu)} K_2 E\alpha |\Delta T_1|$$

$$+ K_3 C_3 E_{ab}|\alpha_a T_a - \alpha_b T_b| + \frac{1}{1-\nu} E\alpha |\Delta T_2| \quad \text{(SI)}$$

(4.17)

where K_1, K_2, K_3 = local stress indices for component under investigation
ν = Poisson's ratio of material
$|\Delta T_1|$ = absolute value of range of temperature difference between temperature of outside surface and inside surface of pipe wall, assuming moment generating equivalent linear temperature distribution (see Fig. 4.5), °F (°C)
$|\Delta T_2|$ = absolute value of range of that portion of nonlinear thermal gradient through the wall thickness not included in ΔT_1 (see Fig. 4.5), °F (°C)

For each S_p, an alternating stress intensity must be calculated:

$$S_{alt} = \frac{K_e S_p}{2} \qquad (4.18)$$

where

$$K_e = \begin{cases} 1.0 & \text{for } S_n \leq 3S_m \\ 1.0 + \frac{1-n}{n(m-1)}\left(\frac{S_n}{3S_m} - 1\right) & \text{for } 3S_m < S_n < 3mS_m \\ \frac{1}{n} & \text{for } S_n \geq 3mS_m \end{cases}$$

m, n = material parameters given in Table 4.4

TABLE 4.4 Material Parameters for Section NB

Material	m	n
Low-alloy steel	2.0	0.2
Martensitic stainless steel	2.0	0.2
Carbon steel	3.0	0.2
Austenitic stainless steel	1.7	0.3
Nickel-chrome-iron	1.7	0.3

SOURCE: Boiler and Pressure Vessle Code Section III, Subsection NB. *(Courtesy of ASME.)*

The alternative stress intensities are used to evaluate the cumulative effect of the stress cycles experienced by the piping system. This is done as follows:

1. The number of times each stress cycle of type 1, 2, 3, etc. is repeated during the life of the system shall be designated as n_1, n_2, n_3, etc. Cycles shall be superimposed such that the maximum possible peak stress ranges are developed.

2. For each type of stress cycle, determine the alternating stress intensity S_{alt} as above.

3. For each value of S_{alt}, use the applicable design fatigue curve to determine the maximum number of cycles permitted if this were the only cycle occurring. (These fatigue curves are found in the appendices to the code.) These values shall be called N_1, N_2, N_3, etc.

4. For each type of stress cycle, calculate the usage factor:

$$U_1 = \frac{n_1}{N_1} \quad U_2 = \frac{n_2}{N_2} \quad U_3 = \frac{n_3}{N_3}$$

5. The cumulative usage factor U is the sum of the individual usage factors:

$$U = U_1 + U_2 + U_3 + \cdots$$

4.4.6 ASME Boiler and Pressure Vessel Code, Section III, Subsections NC and ND

The scope of Subsections NC and ND of the ASME Boiler and Pressure Vessel Code, Section III, encompasses the requirements pertaining to those sections of nuclear piping designated as Class 2 and Class 3, respectively. For discussion of the determinations of piping class, see Chap. 2.

The loadings required to be considered for Subsections NC and ND are the effects of thermal expansion, weight, and other sustained and occasional loads. The equations to be satisfied in order to meet the requirements of Subsections NC and ND are as follows:

Stresses due to sustained loadings. The effects of pressure, weight, and other sustained loads (SL) must meet the requirements of the following equation:

$$S_{SL} = \frac{B_1 PD_o}{2t_n} + \frac{B_2 M_A}{Z} \leq 1.5 S_h \quad \text{(USCS)}$$

$$S_{SL} = \frac{B_1 PD_o}{2t_n} + \frac{1000(B_2 M_A)}{Z} \leq 1.5 S_h \quad \text{(SI)}$$

(4.19)

where B_1, B_2 = primary stress indices for the particular component under investigation
P = internal design gauge pressure, psi (kPa)
D_o = outside diameter of pipe, in (mm)
t_n = nominal wall thickness, in (mm)
M_A = resultant moment loading on cross section due to weight and other sustained loads, in·lb (mm·N)
Z = section modulus of pipe, in³ (mm³)
S_h = basic material allowable stress at design temperature, psi (kPa)

The longitudinal pressure (LP) portion of the above equation and that for occasional loads may be replaced by

$$S_{LP} = \frac{Pd^2}{D_o^2 - d^2} \quad (4.20)$$

where P = design or peak gauge pressure, as applicable, psi (kPa)
d = nominal inside diameter of pipe, in (mm)

Stresses due to occasional loadings. The effects of pressure, weight, other sustained loads, and occasional loads (OL), including earthquake, must meet the requirements of the following equation:

$$S_{OL} = \frac{B_1 P_{max} D_o}{2t_n} + \frac{B_2(M_A + M_B)}{Z} \leq 1.8 S_h \quad \text{(USCS)}$$

$$S_{OL} = \frac{B_1 P_{max} D_o}{2t_n} + \frac{1000 B_2(M_A + M_B)}{Z} \leq 1.8 S_h \quad \text{(SI)}$$

(4.21)

where P_{max} = peak gauge pressure, psi (kPa)
M_B = resultant moment loading on cross section due to occasional loads, such as thrusts from relief or safety valves, loads from pressure and flow transients, and earthquakes, if required; effects of anchor displacement due to earthquake may be excluded if they are included under expansion; in·lb (mm·N)

Equation 4.21 is applicable to those loading conditions designated as service level B, or upset. For service levels C and D (emergency and faulted, respectively), the value $1.2S_h$ should be replaced by $1.8S_h$ and $2.4S_h$, respectively.

Stresses due to thermal expansion. The effects of thermal expansion (TE) must meet either Eq. (4.22) or Eq. (4.23):

$$S_E = \frac{iM_C}{Z} \leq S_A \quad \text{(USCS)}$$

$$S_E = \frac{1000iM_C}{Z} \leq S_A \quad \text{(SI)} \tag{4.22}$$

$$S_{TE} = \frac{PD_o}{4t_n} + \frac{0.75iM_A}{Z} + \frac{iM_C}{Z} + S_h + S_A \quad \text{(USCS)}$$

$$S_{TE} = \frac{PD_o}{4t_n} + \frac{1000(0.75iM_A)}{Z} + \frac{1000iM_C}{Z} \leq S_h + S_A \quad \text{(SI)} \tag{4.23}$$

where M_C = range of resultant moment due to thermal expansion (also moment effects of anchor displacements due to earthquake if not included under occasional loadings), in·lb (mm·N)

S_A = allowable stress range for expansion stresses = $f(1.25S_c + 0.25S_h)$, psi (kPa)

S_c = basic material allowable stress at minimum temperature, psi (kPa)

f = stress range reduction factor for cyclic conditions, as per Table 4.1

Stresses due to unrepeated anchor movement. The effects of an unrepeated anchor movement (AM) (such as building settlement) must meet the following requirement:

$$S_{AM} = \frac{iM_D}{Z} \leq 3S_c \quad \text{(USCS)}$$

$$S_{AM} = \frac{1000iM_D}{Z} \leq 3S_c \quad \text{(SI)} \tag{4.24}$$

where M_D = resultant moment due to unrepeated anchor movement.

The basic material allowable stresses to be used for S_c and S_h for most materials are determined by selecting the minimum of: (1) 0.25 times the ultimate tensile strength at temperature, (2) 0.25 times the ultimate tensile strength at installation temperature, (3) 0.625 times the yield strength at temperature, and (4) 0.625 times the yield strength at installation temperature. These values for approved materials are tabulated in the appendices to the codes.

Subsections NC and ND require use of both stress indices, as used in Subsection NB, and a common stress intensification factor for both in-plane and out-of-plane bending moments as well as for torsion. The formulas for these factors are given in appendices to the code and in Table 4.2.

Chapter 5
Piping Design Loads

5.1 Introduction

The purpose of pipe stress analysis is to ensure the safe operation of piping systems by verifying their structural and pressure-retaining integrity under the loading conditions postulated to occur during the lifetime of the piping in the plant. This is accomplished by calculation of, and comparison to permissible values of, stress in the pipe wall, piping expansion movements, equipment nozzle loads, and system natural frequencies. Additionally, the stress analyst is responsible for determining the design loads for the support so that the system may be restrained safely.

The requirements for stress analysis of piping systems are laid out in the piping codes, a system of rules and standards setting minimum requirements for safe design, construction, and operation of an engineered facility. In the piping industry, the governing code is actually a series of codes administered under the authorization of ANSI and ASME. The various codes, their requirements, and the piping falling under their jurisdiction are discussed in Chap. 4.

Piping design is typically organized into two broad categories: overall system design and detailed component design. The *system* is defined by the fluid distribution system (piping) as a whole, including all in-line equipment (vessels, pumps, and valves). Components include the in-line equipment as well as the piping supports.

The system analysis is used to provide the input to the component analysis in the form of equipment nozzle loads and support loads. In this way, system analysis is usually separated from the component analysis. In many instances, the separation is necessary because different organizations within the project team perform the component analysis as compared to the system analysis. System analysis is normally done by the piping design engineer, while component analysis is usually done by the piping component designer or manufacturer. The component designer, therefore, must work with the component specification and design loads provided by the piping engineer. It is recommended that one organization maintain overall responsibility owing to the interdependency of piping and supports.

Piping systems are typically divided into two main categories and then divided again in subcategories. Basically, pipelines are usually divided into hot and cold systems. *Hot* lines are defined here as those that have a design temperature in excess of 150°F (66°C). The fundamental reason for this division is that hot lines must potentially undergo a flexibility analysis to determine thermal forces, stresses, and displacements.

The hot and cold systems are further subdivided into large- and small-bore (diameter) lines. Typically, those lines with nominal diameters of 2 in (50 mm) and under are classified as small; those with diameters of 2.5 in (65 mm) and greater are large.

The third category by which piping systems may be divided is by code. Certain codes require more rigorous analyses than others, depending on the degree of hazard associated with failure. For example, a hot, large-bore, nuclear Class 1 piping system would probably require a sophisticated computerized analysis, probably with a modal or time history dynamic analysis, whereas a cold, small-bore system designed to B31.8 (gas transmission) code would probably be restrained by using chart methods or engineering judgment.

All the piping codes developed from the same original code—the ASA B31 Code for Pressure Piping, published in June 1935. Because of this common ancestry, many features are shared by these codes. One of these features is the classification of loads into three types: *sustained loads,* or those due to forces present throughout normal operation; *occasional loads,* or those due to forces present at rare intervals during operation; and *expansion loads,* or those due to displacements of pipe. Examples of types of loadings falling within each of the three categories are weight and pressure (sustained); wind, seismic load, vibration, pipe rupture, relief valve discharge (occasional); and pipe thermal expansion, settlement, and differential anchor displacement owing to seismic or thermal equipment movements (expansion).

The loads imposed on the pipe must, in all cases, be transmitted from the pipe to the building structure. This is done by use of pipe supports and restraints. To ensure that the supports and restraints are capable of transmitting the loads, it is necessary to calculate the loads generated by the conditions listed above and to combine them to arrive at a design load.

The methods used to determine pipe support loadings under various conditions vary from simple hand calculations to sophisticated computerized calculations, depending on the complexity of the problem and the criticalness of the pipeline. Calculations for each of the above listed loading types are discussed in this chapter as well as methods for combining these loads to determine a support design load.

5.2 Sustained Loads

Sustained loads, as previously noted, are classified as those loads caused by mechanical forces which are present throughout the normal operation of the piping system. These loads include both weight and pressure loadings.

5.2.1 Sustained loads—weight

All piping systems must be designed for weight loading. Most piping systems can be described as irregular space frames which usually are not self-supporting; therefore, they must be provided with supports to prevent collapse. The supports must be capable of holding the entire weight of the system, including that of the pipe, insulation, fluid, components, and the supports themselves.

The simplest method of estimating pipe stresses and support loads due to weight is to model the pipe as a beam, loaded uniformly along its length. This is especially suitable in cases where the pipe travels in continuous horizontal runs, with a minimum of in-line components or geometry changes. The length of the beam is equal to the distance between supports.

There are two possible ways to model the pipe, depending on the end conditions assumed—the simply supported (pinned-end) beam or the fixed-end beam. For a simply supported beam, the maximum stress and support loads are

$$\sigma = \frac{WL^2}{8Z} \tag{5.1}$$

$$F = \frac{WL}{2} \tag{5.2}$$

where σ = bending stress, psi (N/mm^2)
W = weight per linear unit of pipe, lb/in (N/mm)
L = length of pipe, in (mm)
F = force on support, lb (N)
Z = section modulus of pipe, in^3 (mm^3)

For a fixed-end beam, the maximum stress and support loads are

$$\sigma = \frac{WL^2}{12Z} \tag{5.3}$$

$$F = \frac{WL}{2} \tag{5.4}$$

For either model, the load on each support is the same, i.e., half of the weight of the pipe suspended between the two supports. However, depending on the model chosen, the stress in the pipe varies. Depending on code-mandated allowable stresses, the beam model chosen could affect the permissible support spacing.

Since most weight supports provide only an upward force reaction, the supports behave as those in the simply supported beam model. However, the pipe at the point of support is not free to rotate fully, since it is partially restrained through its attachment to the piping segment beyond the support. In fact, if all pipe runs were of equal length and equally loaded, segment end rotations would cancel each other, causing the pipe to behave as fixed-end beams. Therefore, the true case lies somewhere between the two beam models. For simplicity's sake, most analysts have adopted a compromise case by which stress is calculated as $\sigma = WL^2/(10Z)$, although the calculation $\sigma = WL^2/(8Z)$ may be used when conservatism is required.

Therefore, when straight runs of horizontal pipe are used, it is relatively simple to determine support spacings and subsequently support loads due to weight loadings. Spacing is determined by

$$L = \sqrt{\frac{10ZS}{W}} \qquad (5.5)$$

where L = maximum allowable span between supports for straight horizontal pipe, in (mm)

S = allowable weight stress (dependent on pipe material, temperature, pressure, and code used), psi (N/mm^2)

with all other terms as previously defined

The force on the support from each side is $F = WL/2$, so the weight load on each support would be equal to

$$F = \frac{(10WZS)^{1/2}}{2} \qquad (5.6)$$

where all terms are as above.

To simplify support spacing calculations, Manufacturers Standardization Society (MSS) Standard Practice SP-69 has provided recommended support spacing for various piping sizes. These spans (see Table 5.1), which have been accepted by ASME and ANSI, have been determined by considering insulated, standard wall-thickness pipe, filled with water, limited to a maximum combined bending and shear stress of 1500 psi (10.3 N/mm^2) and maximum pipe sag of 0.1 in (2.5 mm). When these recommended support spacings are adhered to, the stress levels in the piping system due only to weight loading usually need not be explicitly computed.

TABLE 5.1 Suggested Maximum Span between Supports of Pipe.
Basis: Standard pipe at 750°F (399°C), 1500-psi (10.3-N/mm^2) combined stress or 0.1-in (2.5-mm) sag

Nominal pipe size		Water-filled		Steam-, air-, gas-filled	
in	mm	ft	m	ft	m
1	25	7	2.1	9	2.7
2	50	10	3.0	13	4.0
3	75	12	3.7	15	4.6
4	100	14	4.3	17	5.2
6	150	17	5.2	21	6.4
8	200	19	5.8	24	7.3
12	300	23	7.0	30	9.1
16	400	27	8.2	35	10.7
20	500	30	9.1	39	11.9
24	600	32	9.8	42	12.8

SOURCE: MSS SP-69 (*Courtesy of Manufacturers Standardization Society*).

Since most piping systems are not made strictly of straight horizontal runs, the standard support spacing may not be applied uniformly throughout. Locations of supports should consider the following guidelines:

1. Pipe supports should be located as near as possible to concentrated weights such as valves, flanges, etc. From a pipe stress point of view, the best location for support attachment is directly on the equipment; however, this location is often difficult, because of restrictions caused by attachment hardware, configuration constraints, equipment manufacturers' requirements, or operation and maintenance space needs. When it is necessary to calculate stress loads in the pipe, the equipment is usually modeled as a concentrated force, the effects of which can be superimposed on the uniformly loaded beam.

2. When changes of direction in a horizontal plane occur between the pipe and associated supports, such as with pipe elbows, it is suggested that the spacing be limited to three-fourths of the standard span shown in Table 5.1, to promote stability and reduce eccentric loadings. It is preferred that the supports be located near elbows to reduce moments due to directional changes between supports. Note that supports located directly on piping elbows are not recommended since any attachment will stiffen the elbow and require complex stress evaluations.

3. The standard span does not apply to vertical runs of pipe (risers) since no moment (and therefore no stress, as defined by the piping codes) will develop owing to gravity loads in a riser. Support locations are usually selected, and the number of supports per riser determined, by the pipe length and the desired distribution of pipe weight to building steel at the

various floors. It is recommended that a support be located on the upper half of a riser, to prevent buckling due to compressive forces in the pipe and to prevent instability resulting in overturning of the pipe under its own weight. Guides may be used on long vertical risers to reduce pipe sag and resulting excessive piping deflections. These pipe guides are usually placed in span intervals of twice the nominal horizontal span listed in Table 5.1 and do not carry any pipe deadweight.

4. Support locations should be selected near existing building steel to maximize ease of design and construction and to minimize the supplemental structural materials used to transmit the pipe loading back to the building structure.

In cases where the piping does not run in straight, horizontal spans, support loads may be determined by applying a method called *weight balancing*. This method involves breaking the larger piping system into smaller segments of pipe with supports which may be modeled as free bodies in equilibrium and solved statically. This weight balancing is done by choosing the pipe segments such that the number of supports in the model is equal to the number of forces and moments acting on the pipe segment due to gravity. For example, a straight horizontal pipe run may be subjected to one vertical force and one bending moment about the lateral axis by a gravity load; therefore, two supports are required to make a determinate system. A pipe run that changes direction in a horizontal plane is subjected to a vertical force and two perpendicular moments, and so it would require three hangers before a solution could be found. An example illustrating the technique of weight balancing is presented below, with reference to the piping isometric shown in Fig. 5.1.

Problem 5.1 Figure 5.1 shows a pipeline connecting two equipment nozzles (at points A and H). Note that all pipe in Fig. 5.1 is 12-in (300-mm) nominal diameter standard schedule, filled with water and covered with $4\frac{1}{2}$-in (114-mm) thick insulation; all elbows are long radius; and all valves are 150- psi (1034-kPa) pressure rating gate valves. Points A and H are equipment nozzles acting as anchors. The support designer is faced with the task of supporting this line for weight loads; this involves both locating the supports and calculating their loads.

Following the rules laid out previously for locating weight supports, the design engineer would proceed by first determining the standard span. Referring to Table 5.1, we see that the standard span for a straight 12-in (300-mm) diameter water-filled pipe is 23 ft (7 m). Therefore the distance between any two supports on straight horizontal spans should not exceed 23 ft (7 m) for this system.

Pipe supports should be selected first near concentrated weights—in this case, the two valves. The supports should be as close as possible to the valves, considering installation, inspection, and maintenance requirements. For this example, supports will be selected at points B and F, which are located 1 ft (305 mm) from each valve.

The horizontal pipe run between point B and the riser changes direction in the horizontal plane, so the distance to the next support should not exceed 0.75 times the 23-ft (7-m) standard span, or 17.25 ft (5.26 m). Therefore, the next support should be located at a point convenient to building steel structures that maintains the intermediate span within the reduced allowable. In this case, the support will be located halfway down the z axis run, 17 ft (5.2 m) from point B.

The riser, if supported properly, will transmit an axial force to the horizontal run and thus act as a support at that location. The distance from point C to the riser is less than the standard span, so no additional support need be included on this segment of pipe.

As stated previously, there is generally no span requirement for providing pipe supports on a riser, since little moment, and therefore limited stress, is generated by the weight loading. As noted, the riser should be supported at a point on the upper half of its height. The number of supports required for the riser depends on the capacity of the support hardware and the desired distribution of the weight loading to the building steel structure. In this example, the engineer chooses to use two supports on the riser, located at points D and E.

The distance between the riser, which acts as a support for the lower horizontal straight span, and point F is 22 ft (6.4 m). Therefore, no additional supports are required for this segment because this distance is within the 23 ft (7 m) allowable.

The distance between points F and H in Fig. 5.2 is 23 ft (7 m); however, since there is a change in pipe direction in the horizontal plane, this distance exceeds the reduced permissible span of 17.25 ft (5.26 m). Therefore a support is required and is added at a point close to the elbow and convenient to building steel, in this case point G.

Once the supports have been located on the system, the engineer must estimate the loadings on them. As the initial step, weights of the piping materials and in-line components must be determined and tabulated.

There are tables available and published by various piping, pipe support, and equipment manufacturers. These tables, a sample of which is presented in Table 5.2, list weights of pipes and components according to pipe size.

The types of piping and equipment found in this problem consist of 12-in (300-mm) nominal diameter standard pipe, filled with water, with 4.5 in (114 mm) of

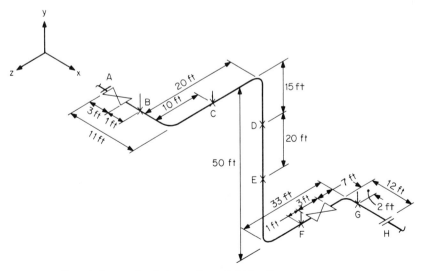

Figure 5.1 Sample isometric of pipe under weight loading.

insulation, 12-in (300-mm) nominal diameter long radius elbows (other information as given above), and 150-psi (1034-kPA) pressure rating gate valves. The weights of these items are determined as below from Table 5.2. The insulation weight was found by interpolating between the values for 4- and 5-in (100- and 125-mm) thickness.

Description	Weight	Insulation	Fluid	Total
12 in (300 mm) standard schedule pipe	49.6 lb/ft (724 N/m)	20.4 lb/ft (298 N/m)	49.0 lb/ft (715 N/m)	119 lb/ft (1737 N/m)
12 in (300 mm) standard schedule long radius elbow	119 lb (530 N)	3 × 20.4 = 61 lb (272 N)	119 lb (530 N)	299 lb (1332 N)
12 in (300 mm) gate valve	925 lb (4116 N)	4.8 × 20.4 = 98 lb (436 N)	3 × 49 = 147 lb (654 N)	1170 lb (5206 N)

The weight of the insulation on the elbow and valve is determined by multiplying the insulation factor (3 and 4.8, respectively) shown below the weights in Table 5.2 by the insulation weight *per foot* of straight pipe. The valve weight should be taken from manufacturer's drawings; however, if this information is not available, it can be estimated from sources such as Table 5.2. Once the actual

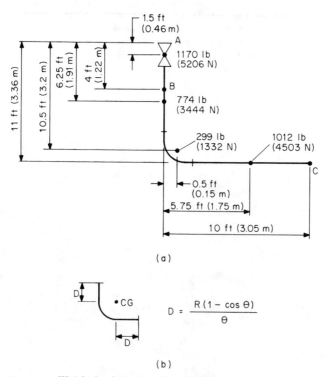

Figure 5.2 Weight loading example—piping segment as free-body diagram: (*a*) plan view; (*b*) center of gravity of elbow.

weight is determined, one need only verify that it is below the assumed value used for the calculations. If not, the previous calculations should be reviewed to determine the impact of the increased loading.

Now that the weight of the pipe and piping components has been estimated, the design engineer may, for analysis purposes, begin to break the piping system into segments representing determinate structures. The first segment (see Fig. 5.1) may be chosen between points A and B (straight horizontal run with two supports) or between points A and C (change of direction in the horizontal plane with three supports). This problem illustrates the latter choice.

Gravity loading on this segment (viewed in plan, Fig. 5.2a) will generate force loadings along the y axis or vertical direction. Support points at the nozzle (point A) and the supports (points B and C) offer resistance to this loading. The centers of gravity of the various piping component weights are as follows:

Valve: 1170 lb (5206 N) at 1.5 ft (0.46 m) from point A

Pipe: 6.5 × 119 = 774 lb (3444 N) at 6.25 ft (1.91 m) from point A

Elbow: 299 lb (1332 N) at 10.5 ft (3.2 m) from point A, 6 in (0.15 m) toward point C

Pipe: 8.5 × 119 = 1012 lb (4503 N) at 5.75 ft (1.75 m) toward point C

Note: The location of the center of gravity of an elbow is determined, as shown in Fig. 5.2b, by the equation

$$D = \frac{R(1 - \cos \theta)}{\theta} \tag{5.7}$$

where D = distance to center of gravity along axis corresponding to axis of either pipe attached to elbow (dimensions consistent with R dimensions)

R = radius of curvature of elbow (dimensions consistent with D dimensions)

θ = angle of bend of elbow, rad

For the example, $R = 1.5$ ft (0.46 m) and $\theta = \pi/2$, so $D = 0.95 \approx 1.0$ ft (0.3 m).

The load on the support at point C can be determined by choosing the x axis to run through points A and B and by summing the moments about that axis. According to the laws of statics, the summation of moments about the x axis will equal zero. The support load at point C can be found as follows:

$\Sigma M_x = 0$

USCS units:

$0 = -299(0.5) - 1012(5.75) + 10C$

$C = 597$ lb (reacting up)

SI units:

$0 = -1332(0.15) - 4503(1.75) + 3.05C$

$C = 2649$ N (up)

Knowing the support load at point C, the engineer can take $\Sigma M_z = 0$ about point A:

$\Sigma M_z = 0$

USCS units:

$0 = -1170(1.5) + 4B - 774(6.25) - 299(10.5) - 1012(11) + 597(11)$

$B = 3574$ lb (up)

TABLE 5.2 Weights of Piping Materials

Pipe or component		Schedule number	12-in pipe size									14-in pipe size									
		Schedule number	30		40		80	100	120	140	160		20	30	40		80	100	120	140	160
		Wall designation		std.		XS								std.		XS					
Pipe		Thickness-in	0.330	0.375	0.406	0.500	0.687	0.843	1.000	1.125	1.312		0.312	0.375	0.437	0.500	0.750	0.937	1.093	1.250	1.406
		Pipe-lb/ft	43.8	49.6	53.5	65.4	88.5	107.2	125.5	139.7	160.3		45.7	54.6	63.4	72.1	106.1	130.7	150.7	170.2	189.1
		Water-lb/ft	49.7	49.0	48.5	47.0	44.0	41.6	39.3	37.5	34.9		60.92	59.7	58.7	57.5	53.2	50.0	47.5	45.0	42.6
Butt welding fittings	Long radius 90° elbow			119 / 3		157 / 3					375 / 3			154 / 3.5		202 / 3.5					
	Short radius 90° elbow			79.5 / 2		104 / 2								102 / 2.3		135 / 2.3					
	45° elbow			60 / 1.3		78 / 1.3					181 / 1.3			77 / 1.5		100 / 1.5					
	Tee			132 / 2.5	(—)	167 / 2.5					360 / 2.5			159 / 2.8		203 / 2.8					
	Lateral			337 / 5.4		556 / 5.4								495 / 5.8		588 / 5.8					
	Reducer			33 / 0.7		44 / 0.7					94 / 0.7			63 / 1.1		83 / 1.1					
	Cap			30 / 1.5		38 / 1.5					89 / 1.5			35 / 1.7		46 / 1.7					

	Temp. range, °F	100 to 199	200 to 299	300 to 399	400 to 499	500 to 599	600 to 699	700 to 799	800 to 999	1000 to 1199	100 to 199	200 to 299	300 to 399	400 to 499	500 to 599	600 to 699	700 to 799	800 to 999	1100 to 1199
Covering	Thickness-in Calcium silicate	1½	1½	2	2½	3	3	3½	4	5	1½	1½	2	2½	3	3	3½	4	5
	Weight lb/ft	6.0	6.0	8.1	10.5	12.7	12.7	15.1	17.9	23.8	6.2	6.2	8.4	10.7	13.1	13.1	15.8	18.5	25.5
	Thickness-in high-temp. comb.						3	3½	4	5						3	3½	4	5
	Thickness-in 85% magnesia	1½	1½	2	2½	3					1½	1½	2	2½	3				
	Total wt/ft	6.0	6.0	8.1	10.5	12.7	17.7	21.9	26.7	35.2	6.2	6.2	8.4	10.7	13.1	18.2	22.8	27.5	37.7
		Cast iron		Steel							Cast iron		Steel						
Flanges	Pressure rating, psi	125	250	150	300	400	600	900	1500	2500	125	250	150	300	400	600	900	1500	2500
	Screwed or slip-on	70 / 1.5	135 / 1.5	72 / 1.5	140 / 1.5	165 / 1.5	250 / 1.5	390 / 1.5	740 / 1.5	1410 / 1.5	93 / 1.5	185 / 1.5	100 / 1.5	195 / 1.5	230 / 1.5	300 / 1.5	480 / 1.5		
	Welding neck			88 / 1.5	165 / 1.5	210 / 1.5	270 / 1.5	390 / 1.5	840 / 1.5	1840 / 1.5			120 / 1.5	210 / 1.5	245 / 1.5	400 / 1.5	480 / 1.5		
	Lap joint			72 / 1.5	165 / 1.5	185 / 1.5	250 / 1.5	440 / 1.5	780 / 1.5	1410 / 1.5			115 / 1.5	220 / 1.5	260 / 1.5	310 / 1.5	495 / 1.5		
	Blind	96 / 1.5	180 / 1.5	120 / 1.5	210 / 1.5	260 / 1.5	345 / 1.5	475 / 1.5	840 / 1.5	1600 / 1.5	126 / 1.5	240 / 1.5	150 / 1.5	280 / 1.5	350 / 1.5	415 / 1.5	600 / 1.5		

TABLE 5.2 Weights of Piping Materials (Continued)

Flanged fittings	Short radius 90° elbow	300 / 5	470 / 5.2	345 / 5	550 / 5.2	700 / 5.5	850 / 5.8	1500 / 6.2		400 / 5.3	620 / 5.5	500 / 5.3	640 / 5.5	670 / 5.7	950 / 5.9	1550 / 6.4
	Long radius 90° ebbow	390 / 6.2	550 / 6.2	480 / 6.2	650 / 6.2			1600 / 6.2		520 / 6.6	770 / 6.6	620 / 6.6	770 / 6.6			
	45° elbow	250 / 4.3	400 / 4.3	280 / 4.3	450 / 4.3	550 / 4.5	725 / 4.7	1130 / 4.8		300 / 4.3	500 / 4.4	380 / 4.3	580 / 4.4	640 / 4.6	880 / 4.8	1250 / 4.9
	Tee	400 / 7.5	670 / 7.8	500 / 7.5	800 / 7.8	950 / 8.3	1300 / 8.7	2000 / 9.3		600 / 8	950 / 8.4	690 / 8	1000 / 8.3	1150 / 8.6	1700 / 8.9	2400 / 9.6
Valves	Flanged bonnet gate	690 / 7.8	1300 / 8.5	925 / 4.8	1350 / 5.5	1600 / 6.8	2275 / 7.1	3250 / 7.8		950 / 7.9	1800 / 8.8	850 / 4.9	1875 / 6.3	2000 / 7.1	3100 / 7.4	4000 / 8.1
	Flanged bonnet globe or angle	800 / 9.4	1200 / 9.5		1400 / 6.5	1500 / 6.8				1175 / 9.9						
	Flanged bonnet check	675 / 9.4	1160 / 9.5	700 / 6.5	875 / 6.5	1100 / 6.8	1175 / 7.1			900 / 9.9						
	Pressure seal bonnet-gate						1700 / 5.2	2100 / 5.5	2500 / 5.9							
	Pressure seal bonnet-globe						1750 / 5.2	2700 / 5.5	3000 / 5.9							
Bolts	One complete flanged joint	15	44	15	49	69	91	124	306	622	22	57	22	62	88	118	159

NOTE: 1 in = 25.4 mm; 1 lb = 4.45 N; 1 psi = 6.895 kPa; temperature °C = (temperature °F − 32)/1.8.
SOURCE: Courtesy of Bergen-Paterson Pipesupport Corp.

SI units:

$$0 = -5206(0.46) + 1.22B - 3444(1.91) - 1332(3.2) - 4503(3.36) + 2649(3.36)$$
$$B = 15,955 \text{ N (up)}$$

The third equilibrium equation $\Sigma F_y = 0$ can be used to find the support load at nozzle A:

$$\Sigma F_y = 0$$

USCS units:

$$0 = A - 1170 + 3574 - 774 - 299 - 1012 + 597$$
$$A = 916 \text{ lb (reacting down)}$$

SI units:

$$0 = A - 5206 + 15,955 - 3444 - 1332 - 4503 + 2649$$
$$A = 4119 \text{ N (down)}$$

The loads calculated at points A and B are the total acting on the supports at these locations. At point C, however, the load calculated represents only that contributed from the segment on one side of the support. To complete the load calculation, it is necessary to continue the weight-balancing operation. The next segment chosen for analysis is that running between points C and D, as shown in Fig. 5.3.

The load on the support at point D can be determined by solving $\Sigma M_x = 0$ about point C:

USCS units:

$$0 = 10D - 1607(10) - 299(9.5) - 1012(4.25)$$
$$D = 2321 \text{ lb (reacting upward)}$$

SI units:

$$0 = 3.05D - 7149(3.05) - 1332(2.9) - 4503(1.3) = 0$$
$$D = 10,334 \text{ N (up)}$$

By setting the sum of the forces acting on the segment equal to zero, the load on the support at point C can be calculated:

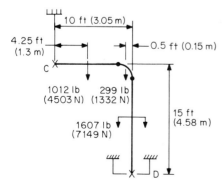

Figure 5.3 Piping segment free-body diagram, elevation view.

USCS units:

$0 = C - 1607 - 299 - 1012 + 2321$

$C = 597$ lb (reacting up)

SI units:

$0 = C - 7149 - 1332 - 4503 + 10{,}334$

$C = 2648$ N (up)

The calculation of the load on the support at point C is again only the partial load on point C; the result must be added to the load calculated from segment A-B-C. Therefore the total reaction of the support at point C is (597 + 597 = 1194 lb) (5296 N) upward (the pipe load on the support acts *downward*).

The calculation of the load on support point D must be completed in a like manner, since segment C-D provides the load from one side. The segment on the far side of point D is a riser section, running between points D and E.

Loadings on multiple supports on a riser cannot be solved by determinate structure analysis, since no moments are developed on the segment. Indeterminate analysis shows that when the two supports are essentially rigid, the weight of the riser between the two supports is divided equally between the two. Thus the contribution to the support at point D is (20 × 119)/2, or 1190 lb (5296 N).

Loads on riser supports may be tailored to suit the engineer's needs through the use of rigid hangers and variable- or constant-spring supports. For example, if the support at point D had been a spring support, preset to a hot (operating) load of 1500 lb (6675 N), the support at point E would be forced to assume an additional load of 1190 + 2321 − 1500 = 2011 lb (8949 N). This would be done if the building steel at point D were not capable of supporting the entire calculated load. Thus, loads may be distributed between supports on a riser without disturbing the balance of the system.

By breaking the system into segments E-F and F-G-H and using the above methodology, loads for each of the supports in Fig. 5.1 may be calculated.

5.2.2 Sustained loads—pressure

Many piping systems are under internal pressure loadings from the fluid they transport. This pressure is usually known for creating stresses in the pipe rather than loadings on the supports. This is because pressure forces are neutralized at the cross section by tension in the pipe wall, as shown in Fig. 5.4, leading to zero net loads on the system as a whole. In Fig. 5.4, the net force at any cross section is equal to $P(A_p) - (PA_p/A_m)A_m = 0$,

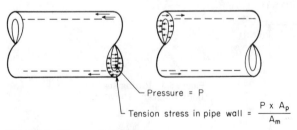

Figure 5.4 Pipe pressure forces.

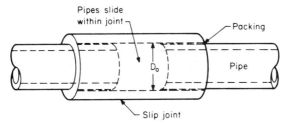

Figure 5.5 Pipe slip joint.

where P = internal pressure, psi (kPa); A_p = internal area of pipe, in² (mm²); and A_m = metal area of pipe, in² (mm²).

When the pipe wall is not continuous from pipe anchor to anchor, the pressure force cannot be counteracted by tension in the pipe wall, and the pressure force must be resisted by restraints and anchors. This occurs when expansion devices (such as slip joints or bellows) are used to absorb movement developed as a result of thermal growth of the pipe.

A slip-type expansion joint (see Fig. 5.5) is simply a telescoping tube with a packing of a sealant material to prevent fluid leakage. The pipe, free to move within the tube, cannot transmit the pressure forces to the joint beyond those forces due to friction of the pipe against the packing.

Another type of expansion device, the bellows fitting (see Fig. 5.6), consists of a series of metal corrugations welded to the pipe to ensure pressure integrity. The bellows looks like and behaves as an accordion, or a helical spring, contracting when a compressive force is applied, and expanding under tension. Since the bellows has a relatively low spring constant as well as a limited strength and load capacity, it does not have the ability to transmit large pressure forces. A restraint is usually required nearby on each side of the joint to prevent the pressure force from pulling apart the expansion joint and rupturing the pressure boundary.

The pressure force developed in the expansion device is equal to the pipe internal pressure times the cross-sectional area over which it is applied. In the case of the slip joint, this is equal to the area covered by the outer dimension of the pipe, or

$$A = \frac{\pi D_o^2}{4} \tag{5.8}$$

where D_o = pipe outer diameter

The equivalent area for a bellows is the maximum cross-sectional area under the corrugation or

$$A = \frac{\pi D_b^2}{4} \tag{5.9}$$

where D_b = maximum internal diameter of corrugation

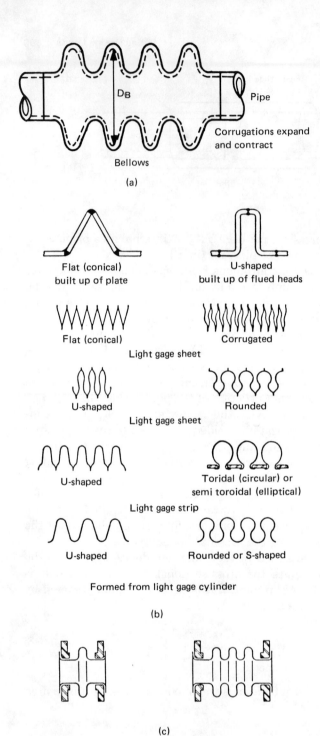

Figure 5.6 Bellows expansion joints: (a) overall view; (b) various shapes of bellows; (c) plain bellows expansion joints.

The dimension of D_b is normally provided by the bellows manufacturer. The force generated by this internal pipe pressure must be absorbed by the closest pipe restraint acting in the direction of application of load.

Problem 5.2 Consider the example illustrated in Fig. 5.7. In this example a 12-in (300-mm) diameter pipe is under an internal gauge pressure of 250 psi (1724 kPa) and has a slip joint located at point C. The pipe is restrained by anchors (full translational and rotational restraints) at points A and E and vertical (y axis) restraints at points B and D.

The force developed due to pressure at the slip joint must be absorbed by restraints on both sides of the joint. The joint force, for a nominal 12-in (300-mm) diameter pipe with internal gauge pressure of 250 psi (1724 kPa) is calculated as follows:

$$F = \frac{P\pi D_o^2}{4} = \frac{(250)\pi(12.75)^2}{4} = 31{,}919 \text{ lb}$$

or

$$F = \frac{(1724)\pi(0.32385)^2}{4} = 142{,}005 \text{ N}$$

On the C-D-E side of the expansion joint, this pressure force is resisted by the restraint at point D. This restraint must be designed to withstand the pressure force in addition to pipe deadweight, thermal forces, and any other loadings to which it will be subjected.

On the A-B-C side of Fig. 5.7, the situation is different in that a moment is developed because of the run of the pipe parallel to the x axis prior to the restraint at B. This pipe segment may be modeled as a beam fixed at one end and pinned at the other with an eccentric loading (as shown in Fig. 5.8).

From beam theory,

$$M_A = \frac{Pb}{2} \qquad F_A = \frac{-3Pb}{2a} \qquad F_b = \frac{2Pa + 3Pb}{2a}$$

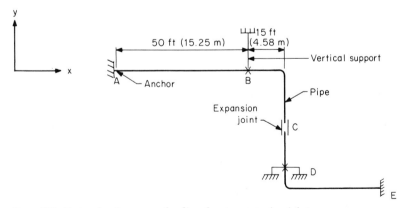

Figure 5.7 Example of pressure loading due to expansion joint.

Therefore the loads on the anchor and restraint can be calculated, with $P = 31{,}919$ lb (142,005 N), $b = 15$ ft (4.58 m), $a = 50$ ft (15.25 m):

$$M \text{ at anchor} = \frac{(31{,}919)(15)}{2} = 239{,}393 \text{ ft·lb}$$

$$= \frac{(142{,}005)(4.58)}{2} = 325{,}191 \text{ m·N}$$

$$F \text{ at anchor} = \frac{(3)(31{,}919)(15)}{(2)(50)} = 14{,}364 \text{ lb}$$

$$= \frac{(3)(142{,}005)(4.58)}{(2)(15.25)} = 63{,}972 \text{ N}$$

$$F \text{ at restraint} = \frac{(2)(31{,}919)(50) + (3)(31{,}919)(15)}{(2)(50)} = 46{,}283 \text{ lb}$$

$$= \frac{(2)(142{,}005)(15.25) + (3)(142{,}005)(4.58)}{(2)(15.25)} = 205{,}977 \text{ N}$$

The vertical restraint must be designed to withstand the load of 46,283 lb (205,997 N) (of uplift, in this case), and the anchor must be capable of resisting the calculated force and moment, in addition to any other design loads.

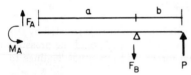

Figure 5.8 Free-body diagram resisting pressure load.

5.3 Occasional Loads

Loads which are applied to a system during only a small portion (typically 1 to 10 percent) of the plant's operating life are usually classified as occasional loads. This classification encompasses loads which vary from periodically applied live loads (such as snow), extreme natural phenomena (hurricane, tornado, earthquake), postulated plant accidents [pipe rupture, loss-of-coolant accident (LOCA), in the case of nuclear plants], and unusual plant operations (relief valve discharge).

Normally occasional loads will subject a piping system to horizontal loads as well as vertical loads, whereas sustained loads will normally be only vertical (weight). It is quite likely that the optimal support locations required for occasional loads will not coincide with those required for sustained loads. It is recommended that, where possible, a little extra conservatism be introduced into the piping design, to permit the use of the same support locations for both sustained and occasional loads. This can be done by reducing the calculated occasional-load spans to coincide with a multiple of the sustained-load spans. The additional cost incurred due to the conservative design will be more than offset by the economy achieved through the use of a single structure for multiple functions.

Dynamic loading is best resisted by rigid supports. However, the system flexibility must be adequate to accommodate thermal growth. When thermal movements are too high to permit the use of rigid restraints, snubbers may be required. Snubbers act as rigid restraints when they are subjected to suddenly applied (dynamic) loads, but snubbers do not resist static loads, such as weight and thermal loads. Where possible, the use of snubbers should be avoided, because of hardware cost and inspection requirements.

The following methods are suggested for determining support locations for occasional loadings:

1. Select initial locations to correspond to deadweight locations, as discussed in Sec. 5.2.
2. Determine the optimal support span for occasional loads by methods explained in this section. Reduce this span until it coincides with a multiple of the deadweight span found in step 1.
3. On cold piping, use rigid supports at all locations.
4. On hot piping, determine locations where rigid supports may be used by methods explained in Sec. 5.3. These locations will normally coincide with locations where "free thermal" movements are minimal. At all other locations, a snubber will probably be required. A number of commercially available computerized optimization programs (such as NPS OPTIM, HANGIT, and QUICK PIPE) may be used to perform this step.

Occasional loadings discussed in this chapter include wind, relief valve discharge, and seismic and vibration loadings. Some piping codes allow for an increase in allowable stress levels of structural materials used for piping systems and supports. This allowable stress may be increased for these temporary loads because they are short in duration and will not hasten creep failure (gradual failure under sustained loads at high temperature) of the piping or supports.

5.3.1 Occasional loads—wind

Wind loading is a periodic force stemming from aerodynamic iteration of the wind or dynamic pressure effects on the piping system. Piping which is located outdoors, and thus exposed to the wind, is normally designed to withstand the maximum wind velocity expected during the operating life of the facility. The magnitude of the wind velocity depends on the local conditions, an estimate of which (for the United States) is shown in Fig. 5.9. This wind velocity, which usually varies with aboveground elevation, is statistically estimated from previous observances.

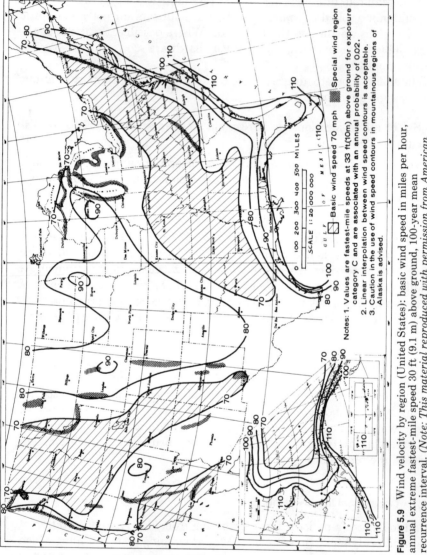

Figure 5.9 Wind velocity by region (United States): basic wind speed in miles per hour, annual extreme fastest-mile speed 30 ft (9.1 m) above ground, 100-year mean recurrence interval. (*Note: This material reproduced with permission from American National Standard A58.1-1982, copyright 1982, copies of which can be purchased from the American National Standards Institute, 1430 Broadway, New York, N.Y. 10018.*)

A large part of the wind loading is caused by the loss of momentum of the wind striking the piping system. Despite localized variations, the force is usually modeled as a uniform load acting over the projected length of the pipe, parallel to the direction of the wind. The expression for the wind force can be adapted from Bernoulli's equation for fluid flow as follows:

$$F = \frac{C_d D q}{386.4} \quad \text{(USCS)}$$

$$F = \frac{C_d D q}{1000} \quad \text{(SI)} \tag{5.10}$$

where F = applied linear dynamic pressure load on projected length, lb/ft (N/m)
C_d = drag coefficient (dimensionless, see Fig. 5.10)
q = dynamic pressure, lb/ft² (N/m²) = $\tfrac{1}{2}\rho V^2$
D = pipe diameter, including insulation, in (mm)
ρ = density of air, lbm/ft³ (kg/m³)
V = velocity of air, ft/s (m/s)

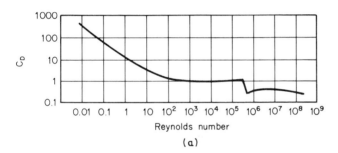

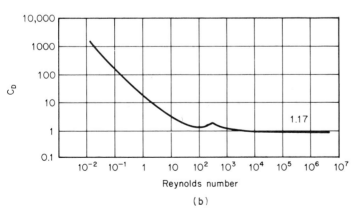

Figure 5.10 Drag coefficients: (*a*) circular cylinders; (*b*) circular and square plates.

The value of the drag coefficient is a function of the shape of the structure and a dimensionless flow factor called the *Reynolds number*. Figure 5.10a and b illustrates the drag coefficient for a cylinder (a pipe) and for flat plates (pipe support components) versus the Reynolds number. (*Note:* Wind loading on the support usually need be considered only in cases where the support has a large surface area projected into the wind.)

The Reynolds number R_n is a dimensionless parameter which gives an indication of degree of turbulence in fluid flow. Low values of the Reynolds number usually indicate steady state or laminar flow while high values indicate transient or turbulent flow of the fluid. This factor is calculated as follows:

$$R_n = \rho \frac{DV}{386.4\mu} \quad \text{(USCS)}$$

$$R_n = \rho \frac{DV}{1000\mu} \quad \text{(SI)}$$

(5.11)

where ρ = air density, lbm/ft³ (kg/m³)
v = wind velocity, ft/s (m/s)
D = pipe diameter or structural length, in (mm)
μ = dynamic viscosity of air, lbf·s/ft² [kg/(m·s)]

From this parameter, the drag coefficient may be estimated from Fig. 5.10a or b.

Under certain conditions, an additional factor of safety, known as a *gust factor*, may be used in calculating the wind load on a piping system. This factor, which usually ranges between 1.0 and 1.3, is used to account for the dynamic effects of non-steady-state air flow. The load calculated previously should be multiplied by this factor when necessary.

Problem 5.3 An example of the principle just described is worked out here to demonstrate the major steps that should be taken as a minimum to design a piping system for outdoor wind loading. (This example does not include the effects of any missile carried by this wind.)

Figure 5.11 depicts an outdoor piping system, of 8-in (200-mm) nominal diameter pipe with 2-in (50.8-mm) insulation, exposed to a postulated maximum wind of 75 mph in the north-south direction. There is a tank connection at nozzle A, and pipe enters the building at point I. The support engineer is required to determine the loads at x direction restraints C, E, and H.

The first step in calculating wind loads on supports is to determine the linear wind load per projected length of pipe:

V = 75 mph = 110 ft/s (33.55 m/s)
ρ_{air} = 0.0748 lbm/ft³ (1.198 kg/m³)   at 29.92 in Hg and 70°F (21°C)
μ_{air} = 39.16 × 10⁻⁸ lbf·s/ft² [1.87 × 10⁻⁵ kg/(m·s)]
D = 8.625 (pipe) + 2 × 2 (insulation) = 12.625 in (320.7 mm)

Next, the Reynolds number is found as follows:

$$R = \frac{(0.0748)(12.625)(110)}{(386.4)(39.16 \times 10^{-8})} = 6.9 \times 10^5$$

or

$$R = \frac{(1.198)(320.7)(33.55)}{(1000)(1.87 \times 10^{-5})} = 6.9 \times 10^5$$

From Fig. 5.10a, C_d is found to be about 0.6 for a Reynolds number of 6.9×10^5. The linear drag force can be calculated, by using a gust factor of 1.3, as

$$F = \frac{(1.3)(0.6)(0.5 \times 0.0748 \times 110^2)(12.625)}{386.4} = 11.5 \text{ lb/ft}$$

or

$$F = \frac{(1.3)(0.6)(0.5 \times 1.198 \times 33.55^2)(320.7)}{1000} = 170 \text{ N/m}$$

This force must now be applied and distributed over the piping system. As stated previously, the wind load is applied to only those portions of the pipe run projecting perpendicular to the wind direction; in this case, that is along any run component parallel to the y or z axis, as shown in Fig. 5.11.

Piping segments parallel to the assumed wind direction such as run segments A-B and G-H will experience minimal wind loading in the x direction. Segment

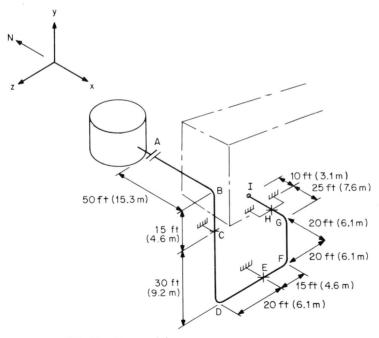

Figure 5.11 Wind loading on piping.

B-D or D-F will see a uniform load in the x direction of 11.5 lb/ft (170 N/m). Segment F-G runs diagonally in the horizontal plane between the x and z directions. Therefore, the projected length of the pipe run is less than the total length of the run, and the F-G segment will see less than the full wind load of 11.5 lb/ft. The actual load can be calculated as

$$\frac{Wl}{L} = \frac{11.5(20)}{\sqrt{20^2 + 20^2}} = 8.1 \text{ lb/ft}$$

$$= \frac{170(6.1)}{\sqrt{6.1^2 + 6.1^2}} = 120 \text{ N/m}$$

where W = wind loading, lb/ft (N/m)
l = projected length, perpendicular to wind direction, ft (m)
L = actual length, ft (m)

This loading is applied to the F-G segment shown in Fig. 5.11 only in the x direction.

The loadings on the supports can be determined by using a method similar to the weight-balancing method. This system can be broken into two segments, as shown in Fig. 5.12, with loads shown at their points of action.

For segment A-E, the equilibrium equations may be used to determine the loads on the restraints. Taking a summation of moments about point A, we get

$\Sigma M_y = 0$

$0 = 20E - (230)(10)$

$E = 115$ lb

or

$0 = 6.1E - (1037)(3.05)$

$E = 519$ N

and

$\Sigma M_z = 0$

$0 = 115(45) - 230(45) - 518(22.5) + 15C$

$C = 1122$ lb

or

$0 = 519(13.8) - 1037(13.8) - 2346(6.9) + 4.6C$

$C = 5073$ N

and

$\Sigma F_x = 115 + 1122 - 230 - 518 + A = 0$

$A = -489$ lb

or

$\Sigma F_x = 519 + 5073 - 1037 - 2346 + A = 0$

$A = -2209$ N

For segment E-H:

$\Sigma M_y = 35H - 229(25) - 172.5(7.5) = 0$

$H = 200.5$ lb (892 N)

$\Sigma F_x = 200.5 - 229 - 172.5 + E = 0$

$E = 201$ lb (894 N)

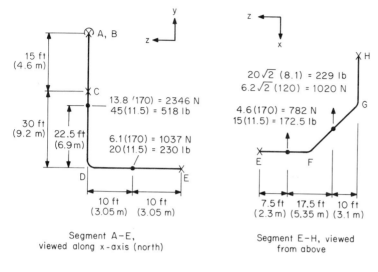

X Restraint location
• CG of wind loading

Figure 5.12 Example of wind loading calculation.

The total load on the restraint at E is the sum of the load from each side, or

$E_{\text{tot}} = 115 + 201 = 316$ lb
$\phantom{E_{\text{tot}}} = 519 + 894 = 1413$ N

5.3.2 Occasional loads—relief valve discharge

Relief valves are used in piping systems to provide an outlet on those occasions when pressure builds up beyond that desired for safe operation. When the pressure setting is reached, the valve opens, allowing sufficient fluid to escape from the piping system to lower the pressure. This permits a controlled discharge of fluid as a means of preventing pressure vessel ruptures.

When a relief valve discharges, the fluid initiates a jet force, which is transferred through the piping system. This force must be resisted by pipe supports if the pipe is not capable of resisting the load internally. The magnitude of the jet force is usually provided by the valve manufacturer. If this value is not known, it may be calculated fairly easily for those cases where the valve vents to the atmosphere. If the fluid discharged flows through a closed system to a vessel, transient flow conditions may develop which make the valve force difficult to calculate.

For a relief valve venting to the atmosphere, the ANSI B31.1 piping code recommends that the discharge force as shown in Fig. 5.13 be calculated as follows:

$$F = \text{DLF}\left(\frac{MV}{32.2} + PA\right) \quad \text{(USCS)}$$

$$F = \text{DLF}\left(MV + \frac{PA}{1 \times 10^6}\right) \quad \text{(SI)} \quad (5.12)$$

where F = discharge force, lb (N)
  DLF = dynamic load factor (dimensionless, see below)
  M = mass flow rate from valve $\times$ 1.11, lbm/s (kg/s)
  V = fluid exit velocity (see below), ft/s (m/s)
  P = static gauge pressure at discharge (see below), psi (N/m²)
  A = discharge flow area, in² (mm²)

Also

$$V = \sqrt{\frac{(50{,}113)(h_o - a)}{2b - 1}} \quad \text{(USCS)}$$

$$V = \sqrt{\frac{(2.0085)(h_o - a)}{2b - 1}} \quad \text{(SI)}$$

where h_o = stagnation enthalpy of pipe fluid, Btu/lbm (J/kg), and a and b are as follows:

Steam condition	a Btu/lbm	J/kg	b (dimensionless)
Wet, <90% quality	291	675,411	11
Saturated, >90% quality	823	1,910,183	4.33
Superheated	831	1,928,751	4.33

And

$$P = \frac{M}{A}\frac{b-1}{b}\sqrt{\frac{48.33(h_o - a)}{2b - 1}} - P_A$$

$$P = \frac{M}{A}\frac{b-1}{b}\sqrt{\frac{(1.995 \times 10^{12})(h_o - a)}{2b - 1}} - P_A$$

where all terms are as before except P_A = atmospheric pressure, psi (N/m²).

The dynamic load factor (DLF) is used to account for the increased load caused by the sudden application of the discharge force. This factor will vary between 1.1 and 2.0, depending on the rigidity of the valve installation and the opening time of the valve. If the piping system is relatively rigidly restrained, the DLF may be calculated by first finding the natural period of vibration of the valve installation:

$$T = 0.1846 \sqrt{\frac{WH^3}{EI}} \quad \text{(USCS)}$$

$$T = 114.59 \sqrt{\frac{WH^3}{EI}} \quad \text{(SI)}$$

(5.13)

where W = mass of safety valve installation, lbm (kg)
H = distance from run pipe to center of outlet pipe, in (mm)
E = pipe material modulus of elasticity at design temperature, psi (N/m²)
I = moment of inertia of inlet pipe, in⁴ (mm⁴)

Next find the ratio of the valve opening t_o to the period T calculated above. For the ratio t_o/T, a DLF can be found from data published in piping codes or structural dynamics texts. A hypothetical DLF curve is shown in Fig. 5.14 for instructional purposes only.

Once the relief valve discharge force has been determined, the load can be distributed to adjacent supports by modeling the pipe segment (and its restraints) as a simple beam.

Problem 5.4 Given a relief valve discharge force of 1500 lb (6675 N) (as specified by the valve manufacturer, including DLF) and the configuration shown in Fig. 5.15, the run pipe at the tee is subjected to the force as well as [due to the 2-ft

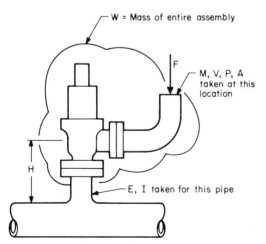

Figure 5.13 Relief valve discharge load calculation.

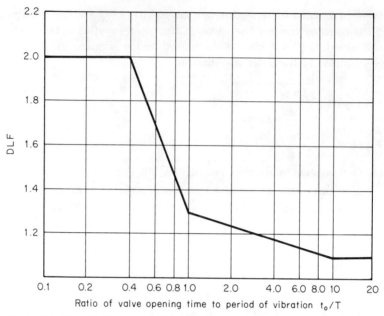

Figure 5.14 Hypothetical dynamic load factor.

(0.6-m) moment arm] a moment of 3000 ft·lb (4072 m·N). The resulting reactions at the restraints can be estimated as

$$F_a = \frac{1500(3)}{17 + 3} + \frac{3000}{20} = 375 \text{ lb}$$

$$F_b = \frac{1500(17)}{17 + 3} - \frac{3000}{20} = 1125 \text{ lb}$$

or

$$F_a = \frac{6675(0.92)}{5.19 + 0.92} + \frac{4072}{6.11} = 1672 \text{ N}$$

$$F_b = \frac{6675(5.19)}{5.19 + 0.92} - \frac{4072}{6.11} = 5003 \text{ N}$$

5.3.3 Occasional loads—seismic

Safety-related piping in nuclear power plants as well as nonnuclear piping in areas where earthquakes are prevalent must usually be designed to withstand seismic loadings.

Earthquake design criteria begin with an estimate of the earthquake potential in an area or region. This earthquake potential is partially based on the known history of previous earthquake activity in the area and is usually determined through a literature search that notes the intensity and the date on which a seismic event may have occurred. The literature search usually consists of a review of reports from old records such as newspapers, journals, etc., and is used to try to estimate the intensity of past earthquakes.

Since seismographs and other instruments capable of measuring earthquake intensity have not been available throughout most of history, the estimations of earthquake intensity must be based on a correlation between reported observations of people witnessing the earthquake and reports of earthquakes of known intensities. An example is shown in Fig. 5.16 which details the expected observations during earthquakes of varying intensity on the modified Mercalli scale.

By performing a search for previous reports and using the data in written documents to estimate the intensity of past earthquakes, a history of a populated location can be assembled and used to help predict further earthquakes in the region. Once the particulars of the earthquake (i.e.,

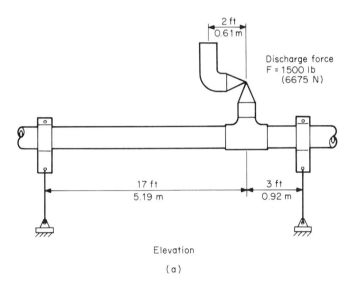

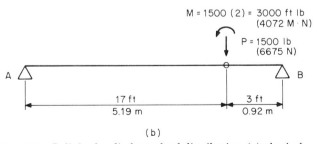

Figure 5.15 Relief valve discharge load distribution: (a) physical configuration; (b) mathematical model.

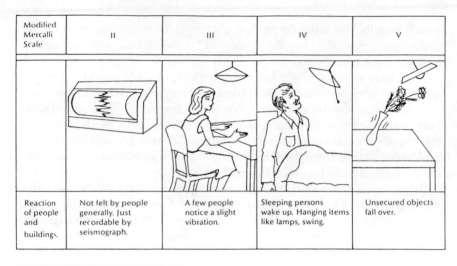

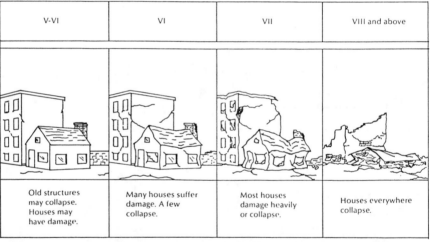

Figure 5.16 Approximate relation between earthquake intensity and observed effects. *(Courtesy of Engineering Planning and Management, Inc.)*

size, duration, etc.) have been established as design criteria, analytical loads to which the structure must be designed can be determined.

Prior to the beginning of the twentieth century, few formal design criteria were used for earthquake design. Buildings in areas of seismic activity were constructed with sufficient space or of such lightweight material that failure of any structure would cause minimal damage to adjacent structures and minimal injury to the inhabitants. Japanese design of housing against earthquakes resulted in lightweight single-story structures.

The Japanese building code was published just prior to the U.S. building code, with both produced because of damage from a severe earthquake. The 1923 earthquake in Tokyo and Yokohama prompted the Japanese to issue their first building code, while the Santa Barbara, California, earthquake of 1927 prompted the United States to do likewise. Since both building codes were developed at about the same time, they have many similarities.

5.3.3.1 Uniform building code. The Uniform Building Code (UBC) is published by an organization that provides data on a national basis. This code provides a means of determining the design loads for earthquakes based on an estimate of the damage potential in a particular region. The damage potential is related to seismic intensity ascribed to four zones labeled 0, 1, 2, and 3, with zone 0 predicting minimal damage and zone 3 predicting major earthquake damage potential. The map showing the various earthquake zones of the United States is shown in Fig. 5.17.

A structure's response to an earthquake varies throughout the event and depends on the frequency and magnitude of acceleration of the ground motion, the natural frequency and damping characteristics of the structure, and the nature of the foundation anchoring the structure to the ground. The building codes have devised methods of converting these dynamic loadings to equivalent static loads. For example, the UBC recommends that a lateral seismic force, assumed to act nonconcurrently along the major axes of the structure, be calculated as

$$V = ZKCW \tag{5.14}$$

where V = lateral seismic force, lb (N)

Z = seismic zone factor: 0.1 for zone 0, 0.25 for zone 1, 0.50 for zone 2, and 1.00 for zone 3

K = building type factor, usually between 0.67 and 3.0

$C = 0.05/T^{1/3}$ but not greater than 0.1

T = fundamental period of structure, s

W = total weight of building, lb (N)

This method may be used to determine seismic loading on commercial piping in those regions covered by the UBC after it is verified that the limiting conditions imposed on the use of this formula have been adhered to.

5.3.3.2 Nuclear seismic design. Prior to 1961 in regions of low seismic activity, such as in the north, south, and midwest of the United States, there was little seismic damage postulated in building designs. The development of the U.S. nuclear power industry and its regulatory bodies, along with the availability of computers, has rapidly expanded earthquake analysis knowledge. The publication of 10CFR100 Appendix A mandated the design of nuclear safety-related structures and systems to withstand earthquake loads.

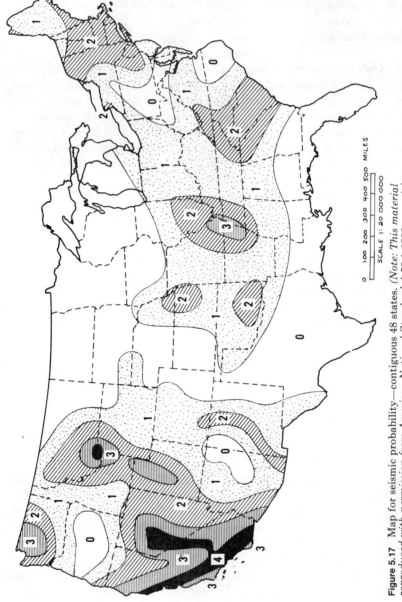

Figure 5.17 Map for seismic probability—contiguous 48 states. (*Note:* This material reproduced with permission from American National Standard A.58.1-1982, copyright 1982, copies of which can be purchased from the American National Standards Institute, 1430 Broadway, New York, N.Y. 10018.)

A commitment by a nuclear plant licensee to comply with adequate earthquake design criteria is made in the safety analysis report (SAR), which is submitted for approval at both the construction permit and operating license stages of regulatory review. According to design criteria established in 10CFR100, plant documentation must substantiate the ability of all seismic category 1 piping systems and components to withstand two levels of site-dependent postulated earthquakes. These are called the *safe shutdown earthquake* (SSE) and the *operating-basis earthquake* (OBE).

The SSE is an earthquake which is postulated based on an evaluation of the maximum earthquake potential given the regional and local geology, seismology, and specific characteristics of local subsurface material. It is that earthquake which produces the maximum vibratory ground motion for which certain structures, systems, and components are designed to remain functional. These structures, systems, and components are those defined in 10CFR100 as necessary to assure (1) the integrity of the reactor coolant pressure boundary, (2) the capability to shut down the reactor and maintain it in a safe shutdown condition, or (3) the capability to prevent or mitigate potential off-site radiation exposure.

The OBE is an earthquake which could reasonably be postulated to affect the plant site during the operating life of the plant. For conservatism, the OBE must usually be equal to at least one-half of the SSE.

Piping may be analyzed for seismic loadings through one of three methods: time history analysis, modal-response spectra analysis, or static analysis.

5.3.3.3 Time history analysis. Time history analysis is based on a record of the postulated earthquake versus time. Data in the form of ground displacement, velocity, or acceleration (as shown in Fig. 5.18) is plotted for the duration of the estimated earthquake record, which may last up to 40 s. This information is plotted for three directions (north-south, east-west, and vertical, or along the major axes of the structure). This data is then used to simulate the seismic excitation of the piping system in a computerized dynamic model, which monitors the stresses, displacements, and restraint loads for the system at regular intervals throughout the seismic event. The computer analysis is usually performed in the elastic region by numerical integration of a lumped-mass mathematical model. The time history computer analysis, although quite accurate, is generally very expensive, since each time interval requires a new calculation.

5.3.3.4 Modal analysis, damping factors, and load combinations. For most applications, the time history method of analysis is too expensive and time-consuming. Therefore, piping systems are frequently analyzed by

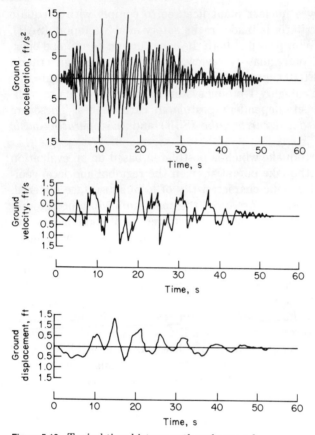

Figure 5.18 Typical time history earthquake record.

modal analysis, using response spectra. Modal analysis breaks the dynamic model of the piping system into a number of modes of vibration (single-degree-of-freedom oscillators), the sum of which approximates the dynamic characteristics of the total system. Response spectra are then generated to determine maximum response (whether acceleration, velocity, or displacement) of idealized single-degree-of-freedom oscillators of varying natural frequencies when they are subjected to the specified input vibratory motion. The maximum response during the seismic event is then determined for each mode of vibration. These responses are combined to determine the total response of the system.

The equation of motion for the single-degree-of-freedom oscillator (see Fig. 5.19) when it is subjected to an externally imposed acceleration is mathematically described as

$$M\ddot{X}(t) + C\dot{X}(t) + KX(t) = Ma(t) \tag{5.15}$$

where M = mass of system, lbm (kg)
 C = coefficient of damping, lbf·s/ft (N·s/m)
 K = spring stiffness, lb/ft (N/m)
 $\ddot{X}(t)$ = response acceleration of mass, ft/s² (m/s²)
 $\dot{X}(t)$ = response velocity of mass, ft/s (m/s)
 $X(t)$ = response displacement of mass, ft (m)
 $a(t)$ = input acceleration, ft/s² (m/s²)

The four last values are functions of time t.

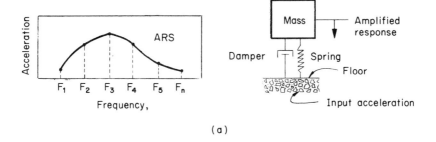

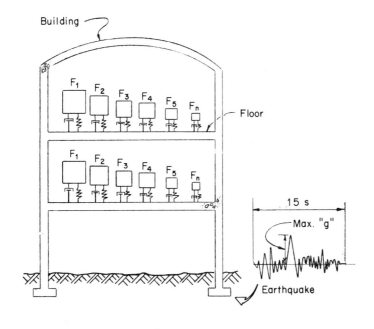

Figure 5.19 Building response spectrum: (*a*) single-frequency excitation; (*b*) multifrequency excitation.

Upon removal of the applied acceleration from an undamped oscillator ($a = 0$, $C = 0$), the system will vibrate at an angular frequency of $\omega_n = \sqrt{K/M}$ rad/s or at $\omega_n/(2\pi)$ Hz. This value is called the *undamped natural frequency* of the system and is used to determine the dynamic response of the oscillator.

Applying a vibratory (harmonic) motion to the above system could develop higher accelerations (and therefore forces) on the mass than had been input at the base, depending on the simultaneous status of the velocity and displacement at the location and time of interest. The acceleration amplification factor can be solved for, and it is found to be

$$\text{Amplification factor} = \sqrt{\frac{1 + (2C_c\,\omega_F/\omega_n)^2}{[1 - (\omega_F/\omega_n)^2]^2 + (2C_c\,\omega_F/\omega_n)^2}} \tag{5.16}$$

where ω_F = forcing angular frequency of input motion, rad/s
ω_n = natural angular frequency of oscillator, rad/s
C_c = ratio of damping coefficient to critical damping, where critical damping = $2\sqrt{KM}$

The forcing frequency of the input motion is determined from the earthquake time history. The *natural frequency* of the oscillator is that at which it will vibrate naturally with no outside stimulus and no damping. *Critical damping* is that level of damping at which the system will no longer oscillate.

The percentage of critical damping to be assumed for piping systems is given by the NRC in Regulatory Guide 1.61, as shown in Table 5.3. Alternative damping values (to those given in Table 5.3 for ASME, Section III, Division I, Classes 1, 2, and 3 piping) are presented in Fig. 5.20 for seismic analysis of piping. The damping values shown in this figure are applicable to both OBE and SSE. This alternative is currently published by ASME as code case N-411, which was approved on September 17, 1984.

By using the above amplification factor, a maximum response may be calculated for single-degree-of-freedom oscillators of all natural frequen-

TABLE 5.3 Damping Values per NRC Regulatory Guide 1.61 as Percentage of Critical Damping

Structure of component	OBE or $\frac{1}{2}$ SSE	SSE
Equipment and large-diameter piping systems, pipe diameter greater than 12 in	2	3
Small-diameter piping systems, diameter equal to or less than 12 in	1	2
Welded steel structure	2	4
Bolted steel structures	4	7
Prestressed concrete structures	2	5
Reinforced concrete structures	4	7

Piping Design Loads   121

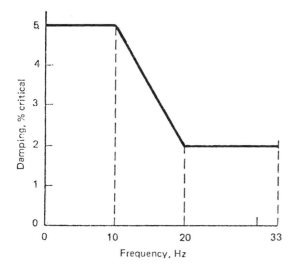

GENERAL NOTE: Applicable to both OBE and SSE, independent of pipe diameter.

Figure 5.20 Damping value for seismic analysis of piping.
SOURCE: Code case N411 (*Courtesy of ASME*).

cies during the course of an earthquake time history. These responses (which need not occur at the same time during the earthquake) are plotted as the response spectrum. This process is shown in Fig. 5.19. A completed response spectrum is shown in Fig. 5.21.

By inspecting the amplification factor, the reason for the characteristic shape of the response spectra becomes evident. When the natural frequency of the oscillator is very small relative to the forcing frequency, the amplification factor approaches zero. This portion of the response spectrum is known as the *flexible range*. As the oscillator frequency approaches the forcing frequencies (most structures transmit strong earthquake motions through their first two modes of vibration), the amplification factor becomes very large, hence the two peaks in the response spectra. This is known as the *resonant range*. When the oscillator frequency is much larger than the forcing frequency, the amplification factor approaches 1.0 and the acceleration becomes the same as the imposed motions. This is known as the *rigid range*, and the acceleration is known as the *zero-period acceleration* (ZPA).

As stated previously, the total response of the piping system is equal to the sum of the responses of the individual modes of vibration. Therefore the seismic loads are dependent on the system natural frequencies.

The piping geometry and restraint configuration affect the system natural frequencies. Therefore these may be altered to "tune" the system to change the piping response and to reduce design loads.

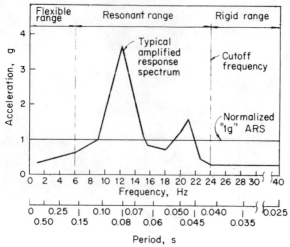

Figure 5.21 Typical response spectrum.

The response spectra essentially represent the building's influence on the piping system during earthquakes. So anything that affects the building natural frequencies, such as a change in the building mass or a building stiffness, will affect the characteristics (i.e., location and magnitude of peaks on the curve) of the response spectra. A perfectly symmetric building such as a square or circular building would probably have the same horizontal response spectra for the north-south and east-west axes. Conversely, an unsymmetric building has separate north-south and east-west response spectra.

The response spectra magnitudes usually increase with the elevation in a building, yet the peak acceleration values occur at approximately the same frequencies. These curve shape characteristics assume there are no major changes in the building design between the various elevations.

Increased damping helps to reduce acceleration values, with a more significant acceleration reduction occurring in the resonant range, as shown in Fig. 5.22. Therefore, it is most important to use an accurate damping value in the calculations. This damping value is a function of internal structural "friction" and is usually best determined through experimental methods. As mentioned previously, NRC Regulatory Guide 1.61 provides damping values which may be used for piping systems when actual values are not known.

When response spectra are used for design purposes, the frequencies at which peak accelerations occur must be given tolerances to account for calculation uncertainties. Figure 5.23 shows how response spectra peaks are usually spread to envelop the uncertainties. The degree of peak

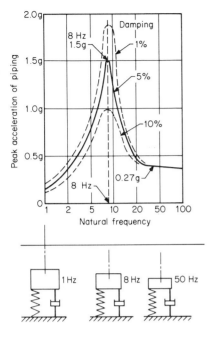

Figure 5.22 Variation of response spectrum with varying damping ratios.

spreading is normally dependent on the confidence level of the design data used to estimate the loading.

As stated previously, the response of the piping system to an earthquake is equal to the sum of the responses of its various modes of vibration. The natural frequencies and mode shapes of the piping system are usually determined by computer analysis of a lumped-mass model, using an eigenvalue-eigenvector (frequency-mode shape) algorithm. However, for preliminary purposes, natural frequencies may be estimated by mod-

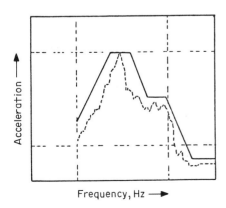

Figure 5.23 Response spectrum peak spreading.

eling the system as a series of connected simple beams. The equation for the natural frequencies of beams with pinned-end conditions is

$$\omega_n = (n\pi)^2 \sqrt{\frac{EI}{ML^4}} \tag{5.17a}$$

$$F_n = \frac{\omega_n}{2\pi} \tag{5.17b}$$

where ω_n = angular natural frequency of nth mode, rad/s
F_n = natural frequency of nth mode, Hz
n = mode of interest
E = modulus of elasticity, psi (kN/m²)
I = moment of inertia of pipe, in⁴ (mm⁴)
M = mass per length of pipe, lbm/in (kg/mm)
L = span between supports, in (mm)

This estimate would lead to error on the low side of the actual frequency and so would give a lower bound for design purposes. If this causes a problem, more exact methods should be used.

Once the natural frequencies for all modes have been determined, the loads generated by the earthquake can be calculated. For systems with multiple node points, the modal response equation would be written as a matrix equation:

$$[F_n] = [M][\phi_n]\Gamma_n a_n \tag{5.18}$$

where $[F_n]$ = force vector for mode n (of forces at each node point), lb (N)
$[M]$ = mass matrix for piping system (of lumped masses at each node point), lbm (kg)
$[\phi_n]$ = shape of mode n normalized to maximum displacement of 1.0 (dimensionless)
Γ_n = mass participation factor for mode n (this is measure of excitability of mode and decreases as mode number increases), dimensionless
a_n = acceleration picked from response spectrum, corresponding to frequency of mode n, ft/s² (m/s²)

Total forces on the system are found by summing the forces for each individual mode. Since the mass participation factor rapidly decreases with higher modes, the contribution of higher modes may usually be neglected.

Two methods for combining loads are the square root of the sum of the square (SRSS) and absolute sum methods, as shown:

SRSS method:

$$[F_t] = \sqrt{[F_1]^2 + [F_2]^2 + [F_3]^2 + \cdots} \tag{5.19}$$

Absolute sum method:

$$[F_t] = [F_1] + [F_2] + [F_3] + \cdots \tag{5.20}$$

The SRSS method is preferred over the absolute sum method because, as noted previously, not all the maximum modal responses occur simultaneously. This method has been endorsed as an acceptable method in NRC Regulatory Guide 1.92. Research has indicated that the SRSS method closely approximates the response from an actual time history analysis in the flexible and resonant range of the response spectra. However, in the rigid range the SRSS method is not a good approximation because there all peak responses will occur simultaneously. Likewise, when modes of a system have closely spaced frequencies (i.e., within 10 percent of each other), the resulting forces must be summed absolutely. This is due to the fact that modes with similar natural frequencies would receive their maximum excitation from the same point during the earthquake time history.

A third summation rule used in some military programs is known as the *Naval Research Laboratory* (NRL) *rule*. This rule requires that the maximum modal response be added absolutely to the SRSS of the remaining modal responses.

Any rule such as SRSS or NRL is based on trial and error, so users of these methods should realize that there could be circumstances in which these rules do not represent an accurate response, especially when significant closely spaced or rigid-range modes are present. Studies have been conducted comparing the time history method to modal analysis using the SRSS summation method. The SRSS results were shown to fall below those of the time history analysis in a few cases, therefore indicating that further study of the use of the SRSS rule may be prudent.

5.3.3.5 Static analysis. Modal analysis, although less expensive and time-consuming than time history analysis, still is complex enough that it must be done on a computer. Where a computer is not available, or where such a level of accuracy is not necessary, earthquake loads on piping may be calculated by static analysis.

One form of static earthquake analysis was described earlier in the discussion of the Uniform Building Code. The formula for earthquake loading presented there yielded a static loading based on the earthquake zone, the natural frequency of the structure, and the building type.

Piping may also be analyzed by static analysis when a response spectrum is available. This may be done when the fundamental (lowest) frequency of the piping system is to the right of the peaks on the response spectra or, better yet, in the rigid range. The acceleration corresponding to this frequency on the response spectra may be used conservatively as the acceleration of the total system. The product of this acceleration and the piping linear weight becomes the uniform load on the pipe.

Problem 5.5 A sample system is shown in Fig. 5.24. The system uses 8-in (200-mm) schedule 40 pipe, not insulated, filled with water, weight = 50.24 lb/ft (733 N/m). Horizontal restraints are spaced typically at 18-ft (5.5-m) intervals. The horizontal acceleration response spectrum applicable to the building and elevation at which this system is located is shown in Fig. 5.25.

The fundamental frequency of the system may be estimated by modeling the pipe spans between restraints as simply supported beams. The lowest frequency, and therefore the frequency of interest, is found where the span is the longest between restraints. In Fig. 5.24, all the restraints are equally spaced. This will rarely occur in reality, so the longest unrestrained span in the direction of interest must be used.

As determined previously in Eqs. (5.17a) and (5.17b),

$$F_n = \frac{(n\pi)^2}{2\pi} \sqrt{\frac{EI}{ML^4}}$$

where F_n = natural frequency of mode = 14.5 Hz
n = number of mode = 1
E = modulus of elasticity = 27.7×10^6 psi (19.1×10^7 kN/m²)
I = moment of inertia of pipe = 72.5 in⁴ (3.02×10^7 mm⁴)
M = mass of pipe per unit length = (50.24 lb/ft)/[(12)(386.4)]
= 0.0108 lbm/in (0.075 kg/mm)
L = length of span = 216 in (5486 mm)

From Fig. 5.25, the corresponding acceleration for a frequency of 14.5 Hz is 2g. This would cause a seismic loading in both the x and z directions of $2 \times 50.24 = 100.5$ lb/ft ($2 \times 0.075 \times 1000 \times 9.8 = 1470$ N/m).

Once the uniform seismic load is applied, the restraint loads may be calculated by a method similar to the one proposed for other types of uniform loads, such as weight or wind loads.

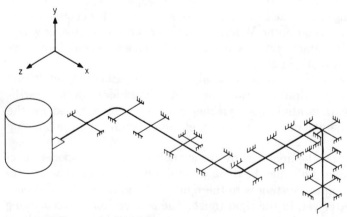

Figure 5.24 System for Problem 5.5.

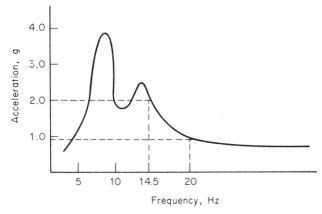

Figure 5.25 Response spectrum.

Often a trade-off may be achieved by tuning the natural frequency of the piping system. For example, the design loads for the system decrease as the fundamental frequency of the system moves from the resonant to the rigid range. This would require more restraints, but the total restraint and pipe loads would be reduced, possibly leading to a material savings. The maximum pipe span between restraints to achieve a response in the rigid range could be calculated by

$$L = \sqrt[4]{\left(\frac{\pi}{2F_r}\right)^2 \left(\frac{EI}{M}\right)} \qquad (5.21)$$

where F_r = frequency designating start of rigid range, Hz.

For Fig. 5.25 the rigid range begins at about 20 Hz, with an acceleration of 0.75g. This would designate a maximum span of 15.3 ft (4.67 m). Therefore, by decreasing the seismic span from 18 to 15.3 ft (5.49 to 4.67 m), the seismic lateral load would be decreased from 100.5 to (0.75)(50.24) = 38 lb/ft (551 N/m).

When no response spectrum is available to indicate the beginning of the rigid range, yet the ZPA is known, a fundamental frequency of 33 Hz is universally accepted as rigid.

5.3.4 Vibration

Rotating equipment such as compressors, pumps, turbine drives, motors, etc., create a large source of mechanical vibration. The rotation creates a harmonic sinusoidal unbalance, providing force which can dynamically excite the piping system.

Unless rotating machinery is very carefully balanced or supported on an elastic foundation designed for vibration prevention, large forces may be generated and transmitted through the piping system. If the rotational frequency is in the neighborhood of the natural frequencies of the piping system, additional resonant vibrations will be induced, leading to possible piping failure.

A compressor of the reciprocating type provides excitation frequencies from periodic pressure sources determined from the rotating speed (rpm) multiplied by the number of cylinders for simple action and twice the number of cylinders for double-action stages. The natural frequencies that should be avoided in restraining piping systems are half of the rpm, the rpm, and all multiples up to 5 times the rpm of the equipment. If these frequencies approach the natural frequencies of the connected piping systems, resonance in the form of large pressure sources can appear. The pressure surge loads can affect the pipes, supports, machinery, and adjacent building structures.

The best way to control vibration and its undesirable effects is to eliminate or isolate the source of vibration. Although elimination and isolation are desirable, in numerous applications these methods are not possible, so some vibration effects should be considered in the piping system design.

5.4 Expansion Loads

We demonstrated that as more restraints are added to a system, the pipe will be more effectively restrained for deadweight and occasional loads. However, most pipe, when operating, increases in temperature and expands. Piping which is too well restrained will not be able to expand, and large forces will develop at the points of lockup, causing large stresses to develop in the pipe.

The ideal restraint condition for thermal considerations is a total lack of restraint. Since this is not feasible, given other loads, some forces due to expansion will develop on restraints even in the most optimally supported system. In this section we show how these loads can be calculated. Additionally we show how to determine piping thermal movements for use in spring hanger selection and design of clearances in restraints.

5.4.1 Determination of thermal loads and stresses

Thermal expansion is of concern primarily parallel to the pipe length axis (axial direction). Thermal expansion can be computed by the following integral:

$$\Delta = L \int_{T_{\text{cold}}}^{T_{\text{hot}}} \alpha \, dT \qquad (5.22)$$

where Δ = thermal expansion in direction specified by length, in (mm)
L = pipe length in direction of interest, in (mm)
α = coefficient of thermal expansion, in/(in·°F) [mm/(mm·°C)]
T = pipe temperature, °F(°C)

For most materials, the coefficient of thermal expansion varies with the temperature, so Eq. (5.22) is not easily applied. Because of this, the thermal expansion in inches per foot between an ambient temperature of 70°F (21°C) and various operating temperatures has been worked out for common piping materials. Values for some of these materials are given in Table 5.4.

The simplest method of calculating thermal loads on supports is by using the guided-cantilever method of modeling the piping system. This method assumes that the thermal growth of the axial runs are absorbed by the bending of pipe legs running perpendicular to the axial growth. Since the displacements force the leg intersections to translate, while rotations are limited by piping continuity, the pipe runs approximate the behavior of guided cantilevers. For a guided cantilever, an imposed displacement induces the following moment and force at each end:

$$M = \frac{6EI\Delta}{L^2} \tag{5.23}$$

$$P = \frac{12EI\Delta}{L^3} \tag{5.24}$$

where P = developed force, lb (N)
M = developed moment, in·lb (mm·N)
E = modulus of elasticity at installed temperature, psi (N/mm^2)
I = moment of inertia of pipe, in^4 (mm^4)
Δ = imposed displacement, in (mm)
L = length of leg perpendicular to direction of growth, in (mm)

The amount of thermal growth absorbed (Δ) by each leg is inversely proportional to the ratio of the stiffness of the subject leg to the sum of the stiffnesses of all legs absorbing the thermal growth. The moments and forces developed must be resisted by the restraint system, whether it be directly by an anchor or by a force couple of two restraints.

Problem 5.6 The system in Fig. 5.26 is made of carbon steel and operates at 350°F (177°C). It uses 12-in (300-mm) standard schedule line with $I = 279$ in^4 (1.16 × 10^8 mm^4) and $E = 27.7 \times 10^6$ psi (1.91 × 10^{11} N/m^2). From Table 5.4, the thermal expansion is found to be 0.0226 in/ft (1.883 mm/m). The system is restrained by two anchors (at points A and G) and two vertical restraints (at points D and E).

The first step in finding the thermal loads is to calculate the pipe expansions and determine the legs which will resist them. The resisting legs are all those perpendicular to the growth, except in those cases where the pipe is restrained at intermediate points. The summary of pipe movements and resisting legs is shown in Table 5.5.

The stiffness of each pipe segment is calculated as $K = 12EI/L^3$. Since the cross-sectional and material properties of all the pipes in this problem are the

TABLE 5.4 Expansion of Pipe, in Inches per Foot from 70°F Base

Temperature °F	Carbon steel carbon moly steel 2¼% Cr – 1% moly	Intermediate alloy steels (5% – 9% Cr – moly)	Austenitic stainless steels (304, 316, 347)	Copper	Brass	Aluminum
−200	0.0180		0.0281	0.0275	0.0287	0.0373
−150	0.0152		0.0236	0.0231	0.0241	0.0310
−100	0.0121		0.0187	0.0183	0.0190	0.0244
−50	0.0087		0.0134	0.0132	0.0137	0.0176
0	0.0051		0.0078	0.0079	0.0081	0.0104
50	0.0015	0.0022	0.0022	0.0023	0.0030	0.0000
70	0.0000	0.0000	0.0000	0.0000	0.0000	0.0000
100	0.0023	0.0022	0.0034	0.0034	0.0035	0.0046
125	0.0042	0.0040	0.0062	0.0063	0.0064	0.0084
150	0.0061	0.0058	0.0090	0.0091	0.0093	0.0123
175	0.0080	0.0076	0.0118	0.0121	0.0123	0.0162
200	0.0099	0.0094	0.0146	0.0151	0.0152	0.0200
225	0.0120	0.0113	0.0175	0.0179	0.0183	0.0242
250	0.0140	0.0132	0.0203	0.0208	0.0214	0.0283
275	0.0161	0.0152	0.0232	0.0238	0.0245	0.0325
300	0.0182	0.0171	0.0261	0.0267	0.0276	0.0366
325	0.0209	0.0191	0.0291	0.0297	0.0308	0.0409
350	0.0226	0.0210	0.0321	0.0327	0.0340	0.0452
375	0.0248	0.0230	0.0351	0.0358	0.0372	0.0496
400	0.0270	0.0250	0.0380	0.0388	0.0405	0.0539
425	0.0293	0.0271	0.0410	0.0419	0.0438	0.0584
450	0.0316	0.0292	0.0440	0.0449	0.0472	0.0628
475	0.0339	0.0314	0.0470	0.0481	0.0506	0.0673
500	0.0362	0.0335	0.0501	0.0512	0.0540	0.0717
525	0.0387	0.0357	0.0531	0.0543	0.0575	0.0764

550	0.0411	0.0379	0.0562	0.0574	0.0610	0.0810
575	0.0436	0.0402	0.0593	0.0607	0.0645	0.0857
600	0.0460	0.0424	0.0624	0.0639	0.0680	0.0903
625	0.0486	0.0447	0.0656	0.0671	0.0717	
650	0.0511	0.0469	0.0687	0.0703	0.0753	
675	0.0537	0.0491	0.0719	0.0736	0.0790	
700	0.0563	0.0514	0.0750	0.0768	0.0826	
725	0.0589	0.0538	0.0783	0.0801	0.0864	
750	0.0616	0.0562	0.0815	0.0834	0.0902	
775	0.0643	0.0586	0.0848	0.0867	0.0940	
800	0.0670	0.0610	0.0880	0.0900	0.0978	
825	0.0697	0.0634	0.0913	0.0934	0.1017	
850	0.0725	0.0658	0.0946	0.0967	0.1056	
875	0.0753	0.0683	0.0979	0.1002	0.1096	
900	0.0781	0.0707	0.1012	0.1037	0.1135	
925	0.0808	0.0732	0.1046	0.1071	0.1173	
950	0.0835	0.0756	0.1080	0.1105	0.1216	
975	0.0862	0.0781	0.1114	0.1140	0.1257	
1000	0.0889	0.0806	0.1148	0.1175	0.1298	
1005	0.0895	0.0811	0.1155			
1010	0.0900	0.0816	0.1162			
1015	0.0906	0.0821	0.1168			
1050	0.0946	0.0855	0.1216			
1100	0.1004	0.0905	0.1284			
1150	0.1057	0.0952	0.1352			
1200	0.1110	0.1000	0.1420			
1250	0.1166	0.1053	0.1488			
1300	0.1222	0.1106	0.1556			
1350	0.1278	0.1155	0.1624			
1400	0.1334	0.1205	0.1692			

NOTE: 1 in = 25.4 mm; 1 ft = 0.305 m; 70°F = 21°C.
SOURCE: Courtesy of Power Piping Co.

same, the relative stiffness of each leg is given by $1/L^3$. Therefore, the proportion of the total displacement absorbed by any given leg n is defined as

$$\Delta_n = \frac{L_n^3}{\Sigma L_i^3} \Delta_T$$

whereas Δ_n = displacement absorbed by leg n, in (mm)
L_n = length of leg n, ft (m)
L_i = length of each other leg resisting specified displacement, ft (m)
Δ_T = total displacement to be absorbed, in (mm)

Once the displacement absorbed by the pipe segment is known, the shear forces and moments can be found by substitution. In this way, the forces and moments on the restraints in Fig. 5.26 can be found:

$$\Delta_X \text{ absorbed by } B\text{-}C = \frac{0.34(30)^3}{30^3 + 60^3 + 30^3} = 0.034 \text{ in}$$

$$= \frac{8.6(9.15)^3}{9.15^3 + 18.3^3 + 9.15^3} = 0.864 \text{ mm}$$

From Eq. (5.24),

$$F_X \text{ across } B\text{-}C = \frac{12(27.7 \times 10^6)(279)(0.034)}{360^3} = 66 \text{ lb}$$

$$= \frac{12(1.91 \times 10^5)(1.16 \times 10^8)(0.864)}{9150^3} = 300 \text{ N}$$

and

$$\Delta_Y \text{ across } A\text{-}B = \frac{0.68(15)^3}{15^3 + 20^3} = 0.202 \text{ in (5.1 mm)}$$

$$F_Y \text{ across } A\text{-}B = \frac{12(27.7 \times 10^6)(279)(0.202)}{180^3} = 3210 \text{ lb (14,285 N)}$$

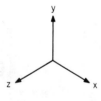
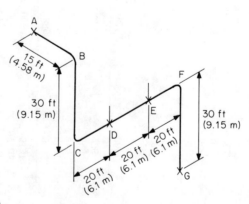

Figure 5.26 Problem 5.6.

TABLE 5.5 Summary of Pipe Movements

From segment	Direction	Magnitude	Resisted by
A-B	X	(0.0226 × 15) = 0.34 in (8.6 mm)	B-C, C-F, F-G
B-C	Y	(0.0226 × 30) = 0.68 in (17.3 mm)	A-B, C-D
C-F	Z	(0.0226 × 60) = 1.36 in (34.5 mm)	A-B, B-C, F-G
F-G	Y	(0.0226 × 30) = 0.68 in (17.3 mm)	E-F

From Eq. (5.23), for segment A-B,

$$M_Z = \frac{6(27.7 \times 10^6)(279)(0.202)}{180^2} = 289{,}096 \text{ in} \cdot \text{lb } (32{,}697 \text{ m} \cdot \text{N})$$

$$\Delta_Z = \frac{1.36(15^3)}{15^3 + 30^3 + 30^3} = 0.08 \text{ in } (2.03 \text{ mm})$$

$$F_Z = \frac{2(27.7 \times 10^6)(279)(0.08)}{180^3} = 1272 \text{ lb } (5661 \text{ N})$$

$$M_Y = \frac{6(27.7 \times 10^6)(279)(0.08)}{180^2} = 114{,}493 \text{ in} \cdot \text{lb } (12{,}950 \text{ m} \cdot \text{N})$$

and

$$M_{X(\text{torsion})@A} = M_{X,B\text{-}C} = \frac{1.36(30^3)}{15^3 + 30^3 + 30^3} \frac{6(27.7 \times 10^6)(279)}{360^2}$$

$$= 228{,}987 \text{ in} \cdot \text{lb } (25{,}899 \text{ m} \cdot \text{N})$$

Therefore the approximate thermal loads on the anchor at point A are

$F_X = 66$ lb (300 N) $\quad M_X = 228{,}987$ in·lb (25,899 m·N)
$F_Y = 3210$ lb (14,285 N) $\quad M_y = 114{,}493$ in·lb (1290 m·N)
$F_z = 1272$ lb (5661 N) $\quad M_z = 289{,}096$ in·lb (32,697 m·N)

Similarly, the loads can be found on the vertical restraints at points D and E:

$$\Delta_{Y,C\text{-}D} = \frac{0.68(20)^3}{15^3 + 20^3} = 0.478 \text{ in } (12.1 \text{ mm})$$

$$F_{Y,C\text{-}D} = \frac{12(27.7 \times 10^6)(279)(0.478)}{240^3} = 3210 \text{ lb } (14{,}285 \text{ N})$$

$$M_{Z,C\text{-}D} = \frac{6(27.7 \times 10^6)(279)(0.478)}{240^2} = 384{,}804 \text{ in} \cdot \text{lb } (43{,}523 \text{ m} \cdot \text{N})$$

$$\Delta_{Y,E\text{-}F} = 0.68 \text{ in } (17.3 \text{ mm})$$

$$F_{Y,E\text{-}F} = \frac{12(27.7 \times 10^6)(279)(0.68)}{240^3} = 4567 \text{ lb } (20{,}321 \text{ N})$$

$$M_{Z,E\text{-}F} = \frac{6(27.7 \times 10^6)(279)(0.68)}{240^2} = 547{,}421 \text{ in} \cdot \text{lb } (61{,}916 \text{ m} \cdot \text{N})$$

The moments at points D and E are resisted by force couples between the two restraints. The total forces can be found by:

$$F_{Y,D} = -3210 - \frac{384{,}805}{240} - \frac{547{,}421}{240} = -7094 \text{ lb } (31{,}570 \text{ N}) \text{ down}$$

$$F_{Y,E} = 4567 + \frac{384{,}805}{240} + \frac{547{,}421}{240} = 8451 \text{ lb } (37{,}608 \text{ N}) \text{ up}$$

Similarly, the forces and moments can be calculated at the anchor at point G.

The guided-cantilever method does not provide fully accurate results because the leg intersection points are actually free to rotate to some degree, thus redistributing the developed moment. In most cases this method yields conservative results: however, in many cases greater accuracy is desired.

Charts have been developed for the purpose of providing the design engineer with accurate load calculations for a wide range of piping configurations. Table 5.6 presents a series of charts developed by the ITT Grinnell Company and reprinted here with their permission. The full series are available in ITT Grinnell's *Piping Design and Engineering*. Sufficient piping configurations are provided to permit the engineer to break down most systems into segments which can be solved easily either by the charts or by the guided-cantilever method.

Most of the Grinnell configurations are of expansion loops. Through judicial use of loops and proper placement of supports, excessive thermal stresses may be reduced. By selecting support locations which direct the expansion to areas of greater flexibility, expansion stresses can be controlled. Expansion loops such as those shown in Fig. 5.27 are used to increase the flexibility of hot piping systems. These loops are designed to accommodate the thermal expansion anticipated during piping operation.

Figure 5.27a illustrates the free thermal movement of the piping, with the terminal locations of piping representing equipment nozzles. The uncontrolled expansion of the pipe can cause locations A and G to become overstressed or overloaded by excessive moments and forces on the nozzles. By providing guides at locations C and F and a limit stop or rigid restraint at location B, as shown in Fig. 5.27b, the thermal expansion can be directed into the expansion loop. Thus the loading on the terminal connections can be reduced significantly.

5.4.2 Determination of Thermal Movements

Piping thermal movements may be estimated at intermediate points in a system by assuming a linear variation between points of known displacement. For example, if the movement of a point located 5 ft (1.5 m) from point A (toward B) in Fig. 5.26 were desired, it could be computed by the following ratios (using the displacements at point B determined in Problem 5.6):

$$\Delta X = \tfrac{5}{15}(0.34) = 0.11 \text{ in } (2.8 \text{ mm})$$

$$\Delta Y = \tfrac{5}{15}(0.202) = 0.067 \text{ in } (1.7 \text{ mm})$$

$$\Delta Z = \tfrac{5}{15}(0.08) = 0.027 \text{ in } (0.69 \text{ mm})$$

These movements are most often required in the y direction for use in spring hanger selection. An example demonstrating the method for determining vertical displacements in a piping system follows.

Piping Design Loads   135

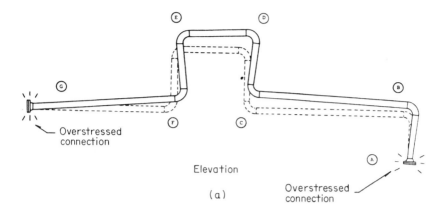

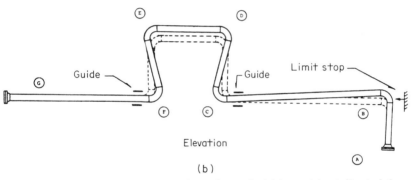

Figure 5.27 Use of supports to direct thermal growth; (a) loop without directed thermal growth; (b) loop with directed thermal movement.

Problem 5.7 The engineer should first generate an isometric drawing such as that in Fig. 5.28, showing all known vertical movements such as nozzle movements of the piping from the cold to hot positions as well as any points of zero vertical movement (due to restraints):

Point A: 2 in (50.8 mm) up, cold to hot
Point C: 0 in
Point F: 4 in (101.6 mm) down, cold to hot
Point K: 1 in (25.4 mm) up, cold to hot
Point L: 0 in
Point M: 0 in

Other sources of vertical movement are the risers B-D and I-J, which expand axially. From Table 5.4 we see that the expansion is $= 0.0707$ in/ft (0.0059 mm/m), so the riser expansions are computed as follows:

$$L_{B\text{-}C} = (0.0707)(15) = 1.06 \text{ in } (26.9 \text{ mm}) \text{ up}$$

$$L_{C\text{-}D} = (0.0707)(30) = 2.12 \text{ in } (53.8 \text{ mm}) \text{ down}$$

$$L_{I\text{-}J} = (0.0707)(10) = 0.707 \text{ in } (18.0 \text{ mm})$$

TABLE 5.6 Piping Expansion Stresses (Expansion factor c)

Temperature, T, °F	Carbon steel, C ≤ 0.30%	Carbon steel, C > 0.30%	C-moly and Low Cr.-moly Cr. ≤ 3%	Cr.-moly 5% ≤ Cr. Mo ≤ 9%	Austenitic stainless steels	Cr. stainless steels 12 Cr., 17 Cr., and 27 Cr.	25 Cr.–20 Ni	Wrought iron
70	0	0	0	0	0	0	0	0
100	37	40	40	35	54	34	47	44
150	98	106	106	92	143	90	125	120
200	160	171	171	149	232	145	204	195
250	228	244	244	212	323	204	287	273
300	294	315	315	271	414	264	368	352
350	305	391	391	335	509	326	455	434
400	436	467	467	396	603	389	541	514
450	510	547	547	465	699	455	629	598
500	584	626	626	531	794	520	716	681
550	664	711	711	603	893	590	809	768

600	743	796	796	672	989	659	901	855
650	827	886	886	714	1089	730	995	946
700	909	974	974	815	1189	799	1088	1035
750	996	1068	1068	891	1292	874	1186	1125
775	1038	1113	1113	929	1344	909	1235	1171
800			1159	967	1395	946	1284	1216
825			1208	1005	1448	983	1335	
850			1256	1043	1500	1022	1384	
875			1303	1081	1552	1061	1435	
900			1351	1121	1605	1097	1484	
925			1398	1161	1659	1134	1533	
950			1445	1200	1713	1174	1585	
975			1492	1240	1766	1212	1634	
1000			1538	1278	1820	1250	1681	
1050			1639	1357	1928	1328	1781	
1100			1737	1435	2036	1404	1879	
1150				1511	2144	1480	1980	

Expansion factor $c = \dfrac{\text{expansion in inches per 100 ft} \times Ec}{1728 \times 100}$

D = outside diameter $\quad I_P$ = moment of inertia of pipe

TABLE 5.6 Piping Expansion Stresses (*Continued*)

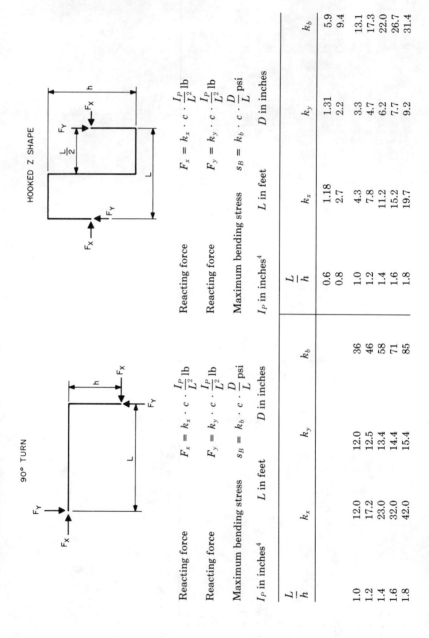

90° TURN

HOOKED Z SHAPE

Reacting force $F_x = k_x \cdot c \cdot \dfrac{I_P}{L^2}$ lb

Reacting force $F_y = k_y \cdot c \cdot \dfrac{I_P}{L^2}$ lb

Maximum bending stress $s_B = k_b \cdot c \cdot \dfrac{D}{L}$ psi

I_P in inches[4] L in feet D in inches

90° TURN

$\dfrac{L}{h}$	k_x	k_y	k_b
1.0	12.0	12.0	36
1.2	17.2	12.5	46
1.4	23.0	13.4	58
1.6	32.0	14.4	71
1.8	42.0	15.4	85

HOOKED Z SHAPE

$\dfrac{L}{h}$	k_x	k_y	k_b
0.6	1.18	1.31	5.9
0.8	2.7	2.2	9.4
1.0	4.3	3.3	13.1
1.2	7.8	4.7	17.3
1.4	11.2	6.2	22.0
1.6	15.2	7.7	26.7
1.8	19.7	9.2	31.4

2.0	54.0	16.6	102	2.0	24.3	10.9	36.4
2.2	68.3	17.8	120	2.2	30.0	13.0	41.5
2.4	84.4	19.2	140	2.4	37.0	15.1	46.6
2.6	103	20.6	161	2.6	45.0	17.2	51.7
2.8	125	22.0	184	2.8	54.0	19.3	58.0
3.0	150	23.5	209	3.0	64.8	21.6	64.8
3.2	175	25.0	234	3.2	76.0	24.0	72.0
3.4	207	26.5	259	3.4	88.0	26.5	79.5
3.6	237	28.0	287	3.6	100.0	29.0	87.0
3.8	274	29.5	318	3.8	113.5	31.6	95.0
4.0	315	31.5	349	4.0	128.7	34.5	103
4.2	356	33.0	381	4.2	144	37.4	112
4.4	406	34.6	414	4.4	160	40.4	121
4.6	456	36.2	450	4.6	178	43.4	130
4.8	510	37.8	487	4.8	198	46.4	139
5.0	570	39.5	528	5.0	219	49.4	148
5.2	630	41.2	569	5.2	241	52.5	157
5.4	700	43.0	610	5.4	263	55.6	167
5.6	775	44.7	652	5.6	286	58.7	176
5.8	855	46.2	696	5.8	310	61.8	185
6.0	938	48.2	743	6.0	334	65.0	195
6.2	1020	49.8	790	6.2	360	68.4	205
6.4	1110	51.6	840	6.4	388	71.8	215
6.6	1212	53.4	892	6.6	416	75.2	226
6.8	1313	55.0	944	6.8	446	78.9	236
7.0	1426	56.8	997	7.0	479	82.0	246
7.2	1517	58.6	1050	7.2	508	85.5	257
7.4	1655	60.2	1104	7.4	540	89.1	268
7.6	1785	61.8	1159	7.6	579	92.7	278
7.8	1917	63.6	1219	7.8	615	96.3	289
8.0	2059	65.4	1284	8.0	653	100.0	300

TABLE 5.6 Piping Expansion Stresses (*Continued*)

Z SHAPE

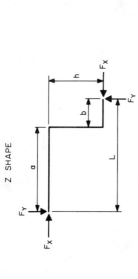

Reacting force $\quad F_x = k_x \cdot c \cdot \dfrac{I_P}{L^2}$ lb

Reacting force $\quad F_y = k_y \cdot c \cdot \dfrac{I_P}{L^2}$ lb

Maximum bending stress $\quad s_B = k_b \cdot c \cdot \dfrac{D}{L}$ psi

I_P in inches^4 $\quad$ L in feet $\quad$ D in inches

$\dfrac{a}{b} \rightarrow$		1			1.5			2			3			4		$\dfrac{a}{b}$
$\dfrac{L}{h}$	k_x	k_y	k_b	k_x	k_y	k_b	k_x	k_y	k_b	k_x	k_y	k_b	k_x	k_y	k_b	$\dfrac{L}{h}$
0.6	9.25	43.0	83.8	8.5	38	91	7.3	32	85	6.5	25	73	6.0	22	66	0.6
0.8	12.8	39.0	69.0	11.8	35	76	10.5	29	71	9.2	23	62	8.5	20	56	0.8
1.0	17.2	37.9	61.9	15.9	34	69	14.4	29	66	12.6	22	52	11.8	19	50	1.0
1.2	22.5	37.8	57.8	21.0	35	69	18	29	66	16.0	23	53	14	20	51	1.2
1.4	28.3	37.7	60.6	27	36	69	22	30	67	20	24	55	19	21	52	1.4
1.6	35.4	42.1	66.3	34	37	71	30	32	69	27	25	57	24	21	55	1.6
1.8	43.0	43.2	72.0	41	39	75	38	33	71	34	26	60	30	22	58	1.8
2.0	52.8	45.7	79.3	50	41	81	46	35	76	42	27	67	40	24	63	2.0
2.2	63.0	48.0	86.5	60	43	88	57	38	83	51	29	73	48	25	68	2.2
2.4	76.0	51.0	93.8	71	46	96	68	40	90	61	31	80	58	27	74	2.4
2.6	89.0	54.5	101.2	83	49	102	79	43	97	71	33	86	69	29	80	2.6
2.8	102	58.2	109.0	96	53	110	91	46	105	82	35	92	80	30	87	2.8

3.0	116	62.2	116.1	110	56	118	104	49	115	92	37	99	90	32	93	3.0
3.2	132	66.0	124.5	124	59	126	118	51	121	106	39	107	104	34	99	3.2
3.4	149	70.0	133.0	140	63	134	133	54	128	121	41	114	118	36	105	3.4
3.6	168	74.0	141.0	168	66	143	149	57	135	136	44	120	132	38	111	3.6
3.8	188	78.0	149.0	177	70	151	165	60	142	151	46	127	147	40	117	3.8
4.0	210	82.0	157.8	197	73	159	181	63	150	166	49	133	163	42	124	4.0
4.2	235	86.4	166.0	219	77	168	201	66	157	185	51	140	182	44	130	4.2
4.4	260	90.6	174.5	241	81	177	221	69	164	204	53	147	201	46	137	4.4
4.6	285	94.8	183.0	263	85	186	241	72	173	223	56	154	220	48	143	4.6
4.8	310	99.0	192.0	287	88	194	263	75	182	243	58	161	239	50	150	4.8
5.0	336	103.2	201.4	314	92	203	288	78	190	264	61	168	260	52	156	5.0
5.2	364	107.6	210.0	341	96	212	313	81	198	286	63	175	281	54	163	5.2
5.4	393	111.8	219.5	370	100	221	339	85	206	310	66	182	304	56	169	5.4
5.6	425	116.2	228.0	399	104	230	365	88	214	335	68	189	329	58	176	5.6
5.8	457	120.5	237.5	430	108	239	392	92	223	360	71	197	355	61	182	5.8
6.0	491	124.8	245.5	461	112	248	422	95	232	386	73	205	381	63	190	6.0
6.2	526	129.4	254.5	493	116	258	450	98	240	414	76	212	408	65	196	6.2
6.4	562	133.8	263.5	526	120	267	478	102	248	443	79	219	436	67	203	6.4
6.6	598	138.2	273.0	561	124	276	506	105	256	472	81	227	465	69	210	6.6
6.8	633	142.6	282.0	598	128	285	535	108	265	502	84	234	495	72	217	6.8
7.0	670	145.0	287.0	636	132	294	565	111	274	533	86	242	526	74	224	7.0
7.2	715	152.0	300.0	674	136	303	601	115	282	565	89	248	557	76	230	7.2
7.4	758	156.5	309.0	714	140	312	639	118	290	599	92	256	588	78	237	7.4
7.6	803	161.0	319.0	756	143	321	680	122	299	633	94	263	620	81	244	7.6
7.8	850	165.5	328.0	798	148	330	724	125	308	668	97	270	655	83	250	7.8
8.0	898	170.0	337.0	840	152	340	770	129	317	703	99	279	694	85	257	8.0

TABLE 5.6 Piping Expansion Stresses (*Continued*)

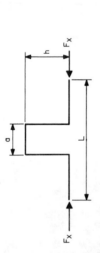

Reacting force $\quad F_x = k_x \cdot c \cdot \dfrac{I_P}{L^2}$

Maximum bending stress $\quad s_B = k_b \cdot c \cdot \dfrac{D}{L}$

I_P in inches[4] $\quad L$ in feet $\quad D$ in inches

$\dfrac{L}{a} \rightarrow$	2		3		4		5		6		7		8		9		10		$\dfrac{L}{a} \rightarrow$
$\dfrac{L}{h}$	k_x	k_b	k_x	k_b	k_x	k_b	k_x	k_b	k_x	k_b	k_x	k_b	k_x	k_b	k_x	k_b	k_x	k_b	$\dfrac{L}{h}$
1.0	2.40	7.20	2.46	8.2	2.52	8.82	2.58	9.29	2.64	9.69	2.67	9.92	2.70	10.1	2.73	10.3	2.75	10.45	1.0
1.2	3.70	9.25	4.46	10.9	4.65	12.0	4.78	12.8	4.84	13.3	5.0	13.9	5.2	14.0	5.29	14.4	5.35	14.9	1.2
1.4	5.31	11.37	6.46	13.6	6.79	15.2	6.98	16.3	7.1	17.0	7.4	17.9	7.7	18.1	7.85	18.6	7.95	19.4	1.4
1.6	7.22	13.53	8.46	16.3	8.93	18.4	9.20	19.8	9.5	20.8	9.8	22.0	10.2	22.3	10.41	22.9	10.55	23.9	1.6
1.8	9.45	15.75	10.48	19.0	11.08	21.6	11.42	23.4	11.9	24.7	12.3	26.1	12.7	26.7	12.97	27.4	13.15	28.5	1.8
2.0	12.00	18.00	12.5	21.8	13.24	24.8	13.87	27.1	14.4	28.8	14.9	30.2	15.3	31.2	15.53	32.3	15.79	33.2	2.0
2.2	14.85	20.25	15.8	24.9	16.6	28.5	16.9	31.0	17.5	33.4	18.0	34.8	18.6	36.2	20.0	38.0	21.0	38.6	2.2
2.4	18.00	22.50	19.6	28.0	20.4	32.2	20.8	35.3	21.3	38.0	22.5	40.0	23.8	41.7	25.2	43.7	26.3	44.4	2.4
2.6	21.52	24.83	23.4	31.1	24.4	35.9	25.5	39.7	26.2	42.7	27.5	45.3	29.0	47.3	30.7	49.5	31.7	50.5	2.6
2.8	25.32	27.10	27.3	34.2	28.9	39.7	30.6	44.0	31.7	47.5	33.0	50.7	34.7	53.0	36.3	55.4	37.2	56.9	2.8
3.0	29.45	29.45	31.2	37.4	33.6	43.7	35.8	48.7	37.7	52.7	39.3	56.2	40.7	59.0	41.9	61.5	43.0	63.6	3.0
3.2	33.9	31.8	35.6	40.6	39.0	47.6	41.2	53.3	43.7	58.0	45.7	61.7	48.0	65.1	50.0	67.8	50.8	70.6	3.2
3.4	38.7	34.1	40.0	43.8	44.5	51.6	46.9	58.0	49.5	63.3	52.2	67.3	55.5	71.3	58.3	74.2	59.2	77.7	3.4
3.6	43.7	36.5	46.1	47.0	50.3	55.6	53.0	62.8	57.5	68.7	59.5	73.0	63.2	77.7	66.7	80.7	68.0	84.9	3.6
3.8	49.1	38.8	52.3	50.2	57.0	59.8	60.2	67.6	65.5	74.1	68.5	79.4	71.5	84.2	75.2	87.3	77.1	92.2	3.8

4.0	54.9	41.1	58.5	53.6	64.0	64.0	69.1	72.5	73.6	79.7	77.5	85.8	80.9	91.0	84.2	94.4	86.6	99.5	4.0
4.2	60.8	43.4	64.7	57.0	71.1	68.2	78.1	77.5	82.0	85.2	87.0	92.2	90.4	97.8	95.0	102.0	97.0	107.0	4.2
4.4	67.3	45.9	71.0	60.4	78.9	72.4	87.2	82.5	91.0	90.8	96.5	98.6	100.5	104.7	106.2	109.6	108.0	114.7	4.4
4.6	73.9	48.2	79.1	63.8	87.0	76.6	96.3	87.5	101.7	96.3	106.5	105.0	112.0	111.7	117.8	117.2	120.0	122.8	4.6
4.8	81.0	50.6	87.2	67.3	95.8	80.8	105.4	92.5	112.4	101.9	118.0	111.4	124.2	118.7	129.8	125.0	133.3	131.0	4.8
5.0	88.2	52.9	95.3	70.8	104.6	85.2	114.7	97.8	122.5	107.5	130.0	117.8	136.7	125.9	142.5	133.0	147.9	139.4	5.0
5.2	95.9	55.3	104.4	74.3	114.0	89.5	125.0	103.0	134.0	113.7	142.0	124.5	149.4	133.2	157.5	141.0	163.0	147.9	5.2
5.4	103.8	57.7	113.5	77.8	123.6	93.9	136.3	108.3	146.0	120.0	155.0	131.3	162.4	140.6	172.6	149.0	178.5	156.5	5.4
5.6	112.1	60.1	122.6	81.3	134.0	98.3	147.6	113.5	159.0	126.2	169.0	138.1	177.0	148.1	187.0	157.1	194.5	165.2	5.6
5.8	120.7	62.4	132.0	84.8	144.6	102.7	159.0	118.8	172.0	132.5	183.0	144.9	192.6	155.6	202.7	165.2	211.0	173.9	5.8
6.0	129.6	64.8	141.6	88.4	155.8	107.0	171.3	124.1	185.2	138.8	197.8	151.8	209.0	163.3	219.2	173.5	228.3	182.6	6.0
6.2	138.8	67.2	152.4	91.9	167.2	111.5	184.0	129.5	199.0	145.1	213.0	158.8	225.0	171.0	236.0	181.8	245.8	191.4	6.2
6.4	148.4	69.6	163.3	95.4	179.1	116.0	198.0	134.9	213.0	151.4	228.5	165.8	241.8	178.8	253.0	190.2	263.8	200.2	6.4
6.6	158.2	71.9	174.2	98.9	191.0	120.5	212.2	140.3	228.0	157.7	245.0	172.9	259.5	186.6	271.0	198.7	282.8	209.1	6.6
6.8	168.4	74.3	185.2	102.4	204.0	125.0	226.4	145.7	244.2	164.0	262.5	180.0	279.0	194.5	292.0	207.2	305	218.0	6.8
7.0	178.9	76.7	196.3	106.0	217.0	129.4	240.7	151.1	261.8	170.3	280.7	187.1	298.7	202.5	314	216.0	328	227.1	7.0
7.2	189.8	79.1	209.1	109.5	230.5	133.9	256.0	156.6	279.8	176.7	299.5	194.3	319	210.5	336	224.7	351	236.6	7.2
7.4	200.9	81.5	221.9	113.0	244.2	138.4	271.5	162.1	297.8	183.1	319.0	201.6	339	218.5	358	233.5	374	246.3	7.4
7.6	212.4	83.8	234.7	116.5	259.2	142.9	287.5	167.6	316	189.5	339.0	208.9	359	226.5	381	242.3	398	256.0	7.6
7.8	224.2	86.2	247.6	120.0	274.5	147.4	304.3	173.1	334	195.9	359.0	216.2	381	234.6	404	251.1	422	265.8	7.8
8.0	236.2	88.6	260.7	123.5	289.8	152.0	322	178.6	352	202.4	379	223.5	405	242.7	427	260.0	448	275.6	8.0
8.2	248.7	91.0	275.0	127.0	305	156.6	340	184.1	372	208.9	400	231.0	429	250.9	451	268.9	475	285.5	8.2
8.4	261.5	93.4	289.3	130.5	320	161.2	358	189.7	392	215.4	422	238.5	455	259.2	476	277.8	502	295.4	8.4
8.6	274.6	95.8	304	134.0	336	165.8	377	195.3	413	221.9	445	246.0	480	267.5	502	286.7	530	305	8.6
8.8	287.9	98.2	318	137.5	351	170.4	395	200.9	434	228.4	470	253.5	505	275.8	529	295.7	560	315	8.8
9.0	302	100.5	332	141.0	367	175.0	416	206.6	456	235.0	495	261.0	530	284.1	559	305	590	325	9.0
9.2	316	102.9	348	144.5	384	179.6	437	212.4	479	241.6	520	268.6	556	292.4	589	314	620	335	9.2
9.4	330	105.4	365	148.0	402	184.2	458	218.2	503	248.2	545	276.2	584	301	618	323	651	345	9.4
9.6	345	107.7	381	151.5	422	188.8	480	224.0	527	254.8	570	283.8	611	309	648	332	684	355	9.6
9.8	360	110.1	397	155.0	443	193.4	502	229.8	551	261.4	596	291.4	639	317	680	342	717	365	9.8
10.0	375	112.5	414	158.7	466	198.1	525	236.1	575	268.2	624	299.0	666	326	711	351	750	375	10.0

TABLE 5.6 Piping Expansion Stresses (*Continued*)

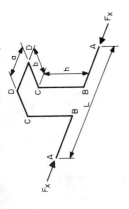

Reacting force $\quad F_x = k_x \cdot c \cdot \dfrac{I_P}{L^2}$ psi

Bending stress $\quad s_B = k_b \cdot c \cdot \dfrac{D}{L}$

Torsional stress $\quad s_t = k_t \cdot c \cdot \dfrac{D}{L}$ psi

I_P in inches $\quad L$ in feet $\quad D$ in inches

Where no value for k_t is listed, the torsional stress is negligible.
NOTE: Letters indicate location of maximum combined stress.

$L/a = 2$

$\dfrac{a}{b} \to$				0.25			0.5			1			2			3			4		$\dfrac{a}{b} \to$
$\dfrac{L}{h}$				k_x	k_b	k_t	k_x	k_b	k_t	k_x	k_b	k_t	k_x	k_b	k_t	k_x	k_b	k_t	k_x	k_b	k_t	$\dfrac{L}{h}$
1				0.20	D 1.63 D		0.64	D 2.79 D	D 0.51 D	1.26	D 2.96 D	D 1.31 D	1.89	A 6.5 A	A	1.91	A 6.6 A	A	2.0	A 6.7 A	A	1
2				0.28	D 2.04 D		1.23	D 4.81 D	D 0.43 D	3.77	D 7.94 D	D 1.72 D	6.95	13.0 A	1.40 A	8.45	15.0 A22.0	1.07 A	9.3	16.0 A	A	2
3				0.32	D 2.15 D		1.53	D 5.62 D	D 0.30	5.70	D 10.6 D	D 1.54 D	13.3	17.0 D	3.15 D	18.2	A 26.4 A	2.76 A	21.0	A 24.5 A	2.32 A	3
4				0.34	D 2.27 D		1.70	D 6.00 D		7.00	D 13.0 D	D 1.41	19.6	19.7 D	5.15 D	28.8	A 26.4 A	4.86 A	35.6	A 31.4 A	4.40 A	4
5				0.35	D 2.32 D		1.81	D 6.25 D		7.86	D 14.3 D		24.4	23.0 D	4.97 D	40.0	A 29.6 A	7.20 A	51.0	A 36.5 A	6.76 A	5
6				0.36	D 2.36 D		1.89	D 6.40 D		8.50	D 15.0 D		28.5	26.2 D	4.77 D	50.7	A 31.6 D	9.58 D	66.7	A 40.0 A	9.28 A	6
7				0.37	D 2.38 D		1.95	D 6.50 D		8.96	D 15.6 D		31.6	28.5 D	4.48 D	59.0	A 35.4 D	9.38 D	81.8	A 42.2 A	11.8 A	7
8				0.38	D 2.40		2.00	D 6.60		9.30	D 15.9		34.2	30.4	4.21	66.6	A 40.0	9.20	95.5	A 43.2	14.2	8

a/b																		
9	0.38	2.42 D	D	204	6.65 D	D	9.58	16.0 D	D	36.3	31.6 D	3.94 D	73.4	43.1 D	9.00 D	108	48.1 D	14.3 D
10	0.39	2.43	D	2.08	6.75	D	9.80	16.3	D	38.0	32.6	3.70	79.0	45.9	8.70	119	52.5	14.1

$L/a = 4$

$a/b \rightarrow$	0.25			0.5			1			2			3			4		
L/h	k_x	k_b	k_t	k_x	k_b	k_t	k_x	k_b	k_t	k_x	k_b	k_t	k_x	k_b	k_t	k_x	k_b	k_t
1	0.67	3.20 D	D	1.22	4.35 D	A 0.30	1.67	A 5.2 A	A 0.15	2.0	C 6.3 C		2.15	C 7.0 C		2.23	C 7.4 C	
2	1.35	5.80 D	D	4.30	9.96 D	2.45 D	6.96	11.0 D	6.55	9.3	15.0 C		10.5	17.7 C		11.0	19.0 C	
3	1.70	7.00 D	D	6.23	13.8 D	2.28 D	14.0	16.5 D	D	21.2	24.0 C		24.2	28.5 C		26.6	31.6 C	
4	1.88	7.44 D	D	7.84	16.9 D	2.09 D	21.3	24.5 D	7.40 D	36.2	30.0 D		44.5	39.5 C		48.5	44.0 C	
5	2.01	7.75 D	D	8.94	18.8 D	1.89 D	27.8	31.4 D	7.75 D	52.6	31.0 D	17.3 D	68.4	49.0 C		76.3	56.0 C	
6	2.09	8.00 D	D	9.72	20.2 D	1.69	33.3	37.0 D	7.70 D	69.5	40.5 D	19.2 D	95.5	56.0 D		109	67.0 C	
7	2.15	8.13 D	D	10.3	21.1 D		37.8	41.6 D	7.70 D	85.7	49.4 D	20.3 D	125	63.6 D	6.75 D	145	76.6 C	5.66 C
8	2.17	8.14 D	D	10.7	21.7 D		41.4	45.0 D	7.14 D	100	57.1 D	20.8 D	155	60.2 D	34.5 D	184	85.4 C	7.42 C
9	2.23	8.30 D	D	11.0	22.2 D		44.4	48.6 D	6.80 D	113	64.0 D	21.3 D	186	71.0 D	36.8 D	226	93.5 C	9.35 C
10	2.26	8.36	D	11.3	22.6	1.69	46.8	50.6	6.45	127	71.5	21.3	216	82.8	38.8	269	101	11.7

SOURCE: *Piping Design and Engineering* (Courtesy of ITT Grinnell Corp.).

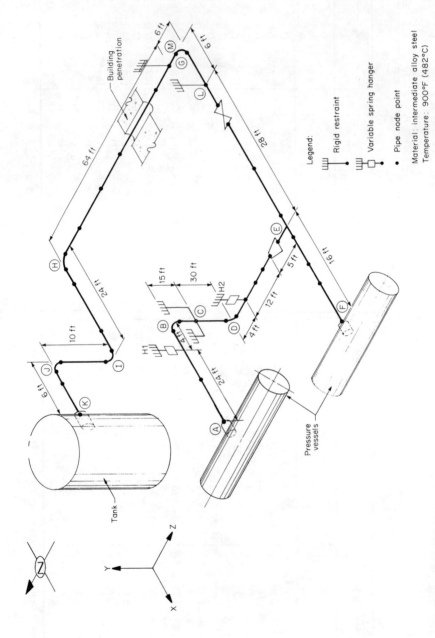

Figure 5.28 Problem 5.7.

The first point of interest is the location of the spring hanger H_1, 4 ft (1.2 m) from point B on pipe segment A-B. The movement of the endpoints has been either a given quantity (as was point A) or computed from thermal growth (as was point B). The thermal movement at intermediate points will be approximately proportional to their distance from the endpoints:

$$H_1 = 1.06 + \tfrac{4}{28}(2 - 1.06) = 1.19 \text{ in } (30.2 \text{ mm}) \text{ up}$$

The vertical movement at H_2 can be determined in the same manner as described above, once the movement at point E is known. The displacement at point E may be calculated, since the displacements are known at endpoints F and L. The displacement at E is found by

$$E = \tfrac{28}{44}(4) = 2.55 \text{ in } (64.8 \text{ mm}) \text{ down}$$

Displacements may now be found at any point between points D and E, since displacements at these points are known [2.12 in (53.8 mm) down and 2.55 in (64.8 mm) down, respectively]. Therefore the displacement at H_2 is $2.12 + \tfrac{4}{21}(2.55 - 2.12) = 2.2$ in (55.9 mm) down.

Total displacement to be absorbed by the horizontal legs between points K and M is the difference between the nozzle movement at K and expansion of riser I-J, or $1 - 0.707 = 0.273$ in (6.9 mm). The total horizontal length between K and M is 94 ft (28.7 m). Therefore, the displacement at point J is approximately $1 - \tfrac{6}{94}(0.273) = 0.983$ in (25.0 mm) up, the displacement at point I is $0.983 - 0.707 = 0.276$ in (7.0 mm) up, the displacement at point M is zero, and the displacement at any intermediate point may be estimated by linear interpolation.

The support engineer may locate rigid restraints as desired to force points of zero displacement, and direct thermal displacement to easily calculated locations. The permissible location is dependent on the displacement to be absorbed by the overhanging leg, the pipe size, and the allowable stress of the pipe. The formula for the minimum permissible distance to the first rigid restraint is

$$L = \sqrt{\frac{\Delta D(10^6)}{1.6 S}} \quad \text{(USCS)}$$

$$L = \sqrt{\frac{\Delta D(6.2 \times 10^5)}{S}} \quad \text{(SI)} \tag{5.25}$$

where L = distance to first rigid restraint, ft (m)
  Δ = displacement to be absorbed, in (mm)
  D = outer diameter of pipe, in (mm)
  S = allowable stress of pipe, psi (N/m²)

Table 5.7 gives distances (in feet) to the first rigid restraint for allowable stresses of 10,000 psi (68,948 kN/m²).

Once the rigid restraints have been located, loads and movements may be calculated as described previously.

TABLE 5.7 Distances to the first rigid restraint for allowable stresses of 10,000 psi $\left(L = \sqrt{\dfrac{\Delta \times \text{OD of pipe} \times 10^6}{1.6S}}\right)$ $S = 10{,}000$ psi

Deflection, in	\multicolumn{17}{c}{Pipe size, in}
	1	$1\tfrac{1}{4}$	$1\tfrac{1}{2}$	2	$2\tfrac{1}{2}$	3	$3\tfrac{1}{2}$	4	5	6	8	10	12	14	16	18	20
$\tfrac{1}{4}$	4.5	5.0	5.5	6.0	6.5	7.5	8.0	8.5	9.5	10	11.5	13	14	15	16	17	17.5
$\tfrac{1}{2}$	6.5	7.0	7.5	8.5	9.5	10.5	11	12	13	14.5	16.5	18.5	20	21	22.5	23.5	25
$\tfrac{3}{4}$	8.0	9.0	9.5	10.5	11.5	13	14	14.5	16	17.5	20	22.5	24.5	25.5	27.5	29	30.5
1	9.0	10	11	12	13.5	15	16	17	18.5	20.5	23	26	28	29.5	31.5	33.5	35.5
$1\tfrac{1}{4}$	10	11.5	12	13.5	15	16.5	17.5	18.5	21	22.5	26	29	31.5	33	35.5	37.5	39.5
$1\tfrac{1}{2}$	11	12.5	13.5	15	16.5	18	19.5	20.5	23	25	28.5	32	34.5	36	38.5	41	43.5
$1\tfrac{3}{4}$	12	13.5	14.5	16	17.5	19.5	21	22	24.5	27	30.5	34	37	39	42	44.5	47
2	13	14.5	15.5	17	19	21	22.5	23.5	26.5	29	33	36.5	40	42	44.5	47.5	50
$2\tfrac{1}{4}$	13.5	15.5	16.5	18.5	20	22	23.5	25	28	30.5	35	39	42.5	44.5	47.5	50.5	53

	14.5	16	17	19.5	21	23.5	25	26.5	29.5	32	36.5	41	44.5	47	50	53	56
2¼	14.5	16	17	19.5	21	23.5	25	26.5	29.5	32	36.5	41	44.5	47	50	53	56
2¾	15	17	18	20	22	24.5	26	28	31	33.5	38.5	43	47	49	52.5	55.5	58.5
3	15.5	17.5	19	21	23	25.5	27.5	29	32.5	35	40	45	49	51	55	58	61
3½	17	19	20.5	23	25	27.5	29.5	31.5	35	38	43.5	48.5	53	55.5	59	63	66
4	18	20.5	22	24.5	27	29.5	31.5	33.5	37.5	40.5	46.5	52	56.5	59	63	67	70.5
4½	19	21.5	23	26	28.5	31.5	33.5	35.5	39.5	43	49.5	55	60	63	67	71	75
5	20.5	23	24.5	27	30	33	35.5	37.5	41.5	45.5	52	58	63	66	70.5	75	79
5½	21.5	24	25.5	28.5	31.5	34.5	37	39.5	43.5	47.5	54.5	61	66	69.5	74	78.5	83
6	22	25	26.5	30	33	36	38.5	41	45.5	50	57	63.5	69	72.5	77.5	82	86.5

Correction Factors for Bending Stresses Other than 10,000 psi

Bending stress	2000	3000	4000	5000	6000	8000	10,000	12,000	15,000	20,000
Multiply L by:	2.24	1.83	1.58	1.41	1.29	1.12	1.00	0.91	0.82	0.71

5.5 Load Combination

Pipe supports must be designed to withstand any combination of loading which is postulated to occur simultaneously. Normal operating loads are either deadweight or deadweight plus thermal. These loads may be combined with occasional loads, as required by the design criteria of the specific project.

Under certain conditions, capacities of materials may be increased for occasional loads. An example is the concept of service level instituted by ASME Boiler and Pressure Vessel Code, Section III (for more information see Chap. 4). A typical specification for design load combinations is as follows:

Normal:	Deadweight
	Deadweight + thermal
Upset:	Normal + relief valve discharge
	Normal + earthquake (OBE) + relief valve discharge
	Normal + water hammer
Emergency:	Normal + earthquake (SSE) + relief valve discharge
Faulted:	Normal + earthquake (SSE) + pipe rupture

These loads may be added either algebraically to arrive at realistic values or absolutely for added conservatism, according to the design criteria requirements.

5.6 Computer Calculations

In this chapter so far we have dealt with manual methods used to estimate pipe support loads. However, most piping work today is done with the assistance of computer pipe stress analysis programs. A large number of programs are available today, varying with respect to capabilities and input/output requirements which reflect their intended industrial applications. However, the differences are mostly external, since all the programs are basically similar. The pipe stress programs are actually structural programs, where the solution is determined by either a stiffness or a flexibility matrix. The piping system is modeled as an irregular space frame and subjected to the various desired loading cases. The supports develop loads as part of the indeterminate analysis; these can be printed out for each case or combined from two or more load cases as desired.

Further information on computerized pipe stress programs, as well as a sample annotated output, can be found in Chap. 9.

Chapter 6

Pipe Support Hardware

6.1 Introduction

For nearly every function of pipe support, several choices of hardware types and sizes may be used depending on the physical constraints of the design and the magnitude of the loads. *Pipe support hardware* refers to the physical structural elements such as preengineered vendor-supplied equipment (struts, springs, clamps, etc.) and structural steel. In this chapter we discuss the various types of hardware normally available from piping supply vendors and their use in conjunction with designer-supplied items such as supplementary steel, concrete anchored baseplates, embedded plates, and welded pipe attachments. Figure 6.1 depicts a number of these items in a plant setting.

The major criteria governing hardware selection are support function, magnitude of the expected load, and space limitations. However, the following must also be kept in mind:

1. The design temperature used for selection of pipe clamps, U-bolts, straps, and other steel in direct contact with the pipe is that of the fluid contents of the pipe. The strength of these items may be expected to decline as the design temperature increases.

The expansion effects of pipe temperature must also be considered in sizing gaps and designing for friction forces.

2. Piping operating at high temperature or subject to condensation on the outer surface will usually be insulated. The pipe support hardware must be designed to accommodate the insulation.

Figure 6.2 illustrates piping insulation, which would normally cover the pipe attachment as well as the pipe. For hostile environments such as in nuclear plant containment areas, the insulation must be resistant to heat, humidity, and radiation. Additionally, it should be quickly and easily removed and reinstalled to allow periodic inspection of piping and equipment.

The thickness of the insulation on the pipe should be determined prior to the selection of the hardware.

3. The piping, attachments, and supporting structure in contact with each other must be of compatible materials in order to reduce galvanic action. In certain cases it may be necessary to insert inert packing material between dissimilar pipe and support materials. Additionally, the support materials must be suitable for the environment to which they will be subjected during the design lifetime of the piping system.

4. All hardware should be periodically inspected, beginning with the start-up of the system. The follow-up inspection schedule is dependent on the criticalness of the system, the service record of the hardware type, and the severity of the service environment. Hardware types requiring less inspection are preferable wherever possible.

The most commonly used support types are listed below. A discussion describing the hardware selection and sizing process for typical pipe loadings follows:

1. Weight supports (rod hangers, sliding supports, variable-spring hangers, constant-spring hangers)

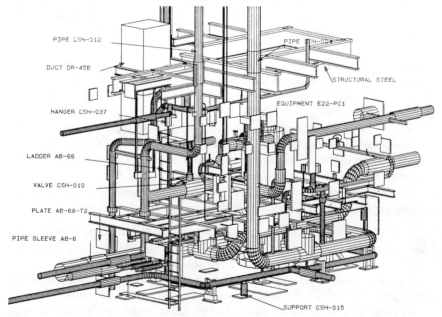

Figure 6.1 Piping runs in power application. *(Courtesy of Construction Systems Associates, Inc., Marietta, Ga.)*

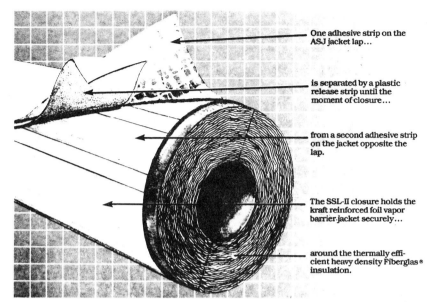

Figure 6.2 Typical insulation for piping. *(Courtesy of Owens-Corning Fiberglas Corp.)*

2. Rigid restraints and anchors (clamps, struts, support steel, welded attachments)
3. Snubbers (hydraulic and mechanical)
4. Sway braces

6.2 Weight Supports

Weight supports are provided to resist vertical pipe loads occurring in the downward (gravity) direction only. They are used on systems for which deadweight is the only design load as well as at support locations where uplifting loads such as thermal and seismic loads are not sufficient to overcome the downward weight load. Hardware items typically used in weight supports are shown in Fig. 6.3.

6.2.1 Rigid supports

The choice of weight support type is made based on the thermal movement expected to occur at the support location during plant operation. When vertical thermal movement is expected to be negligible (such as on a cold line or when thermal movement has been successfully directed elsewhere), adequate support can be provided for nonvibrating pipes by simply resting the pipe on the support. The load is then transmitted back to building steel.

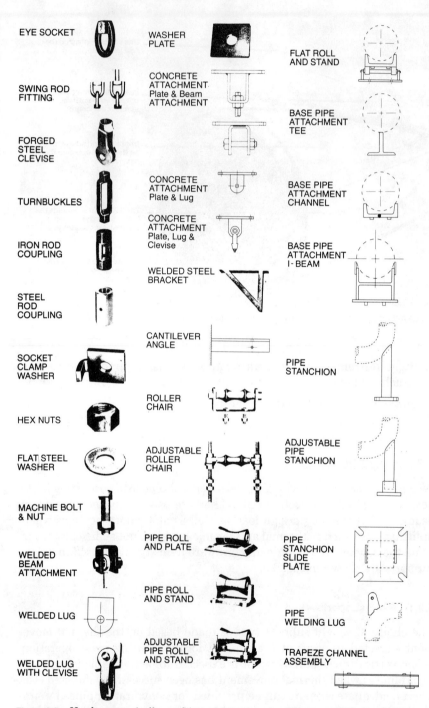

Figure 6.3. Hardware typically used in weight supports. *(Courtesy of Corner & Lada Co., Inc.)*

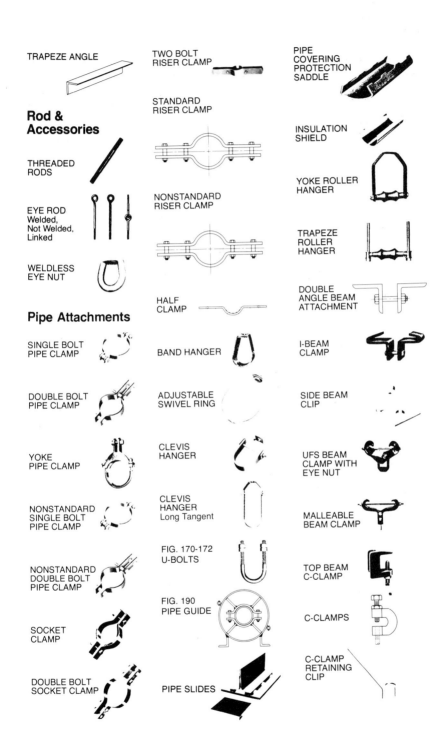

156 Piping and Pipe Support Systems

The most economical and common method of transmitting weight loads to building steel is with rod hanger assemblies. Rods are strong in tension, while their small radius of gyration makes them inadequate for any significant compression loading. Therefore, the rods must be located above the pipe, to hang the pipe from the steel. A number of hardware items, such as those shown in Fig. 6.3, are available for attaching the rod to either the building structure on one side or the pipe on the other. Examples of typical rod assemblies are shown in Fig. 6.4.

Each of the components shown in Figs. 6.3 and 6.4 are supplied by manufacturers in various sizes, with preengineered load ratings, as shown in Table 6.1. These load ratings simplify the design process by providing a prequalified load-rated pipe support. Whenever load-rated hardware is used, analysis steps required in the design of steel may be eliminated, saving costly engineering time. Aside from load capacities the designer must keep some additional considerations in mind when designing a rod hanger support:

1. Horizontal thermal pipe movement must not develop an excessive angle ($\pm 4°$) off the vertical axis during operating condition (see Fig. 6.5a). Large thermal movement both lifts the pipe, introducing additional pipe stresses, and causes horizontal loads that can cause the piping system or

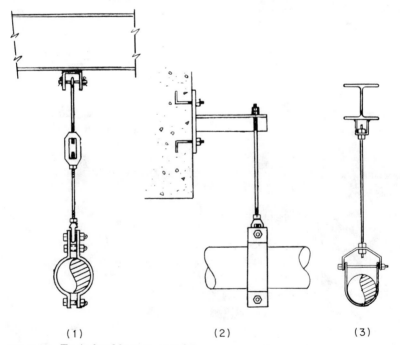

Figure 6.4 Typical rod hanger assemblies.

TABLE 6.1 Load-Carrying Capacities of Hot-Rolled Steel Rod (ASTM Specification A-36)

Nominal rod diameter, in	Root area thread, in^2	Maximum safe load, pounds	
		Rod temperature = 650°F	Rod temperature = 750°F
3/8	0.068	610	540
1/2	0.126	1,130	1,010
5/8	0.202	1,810	1,610
3/4	0.302	2,710	2,420
7/8	0.419	3,770	3,360
1	0.552	4,960	4,420
1 1/4	0.889	8,000	7,140
1 1/2	1.293	11,630	10,370
1 3/4	1.744	15,700	14,000
2	2.300	20,700	18,460
2 1/4	3.023	27,200	24,260
2 1/2	3.719	33,500	29,880
2 3/4	4.619	41,580	37,066
3	5.621	50,580	45,085
3 1/4	6.720	60,480	53,906
3 1/2	7.918	71,280	63,493
3 3/4	9.214	82,890	73,855
4	10.608	95,400	85,001
4 1/4	12.100	109,000	97,119
4 1/2	13.700	123,000	109,593
4 3/4	15.400	138,000	122,958
5	17.200	154,000	137,214

support to become unstable. To limit the angle, the support may have to be offset in the cold position as shown in Fig. 6.5b. Where this is not possible, a sliding support or a roller-type assembly should be used.

2. Rod hangers, though classified as weight supports, at peak load may also carry additional loads such as downward thermal or seismic loads. Thus, the support must be designed for the maximum expected downward load. In trapeze-type assemblies, the rods should be designed to accommodate load redistribution due to possible shifting of the pipe position during operation, by having capacities above 50 percent (that is, 75 percent) of the total load.

3. As a minimum, a 3/8-in (10-mm) diameter rod should be used for small-bore [2-in (50-mm) diameter and under] piping and 1/2-in (12-mm)

diameter rod should be used for large-bore [2.5-in (65-mm) diameter and larger] piping.

4. Rod hanger hardware shown in Fig. 6.4 has some amount of hanger adjustability. The rod and bolt assemblies shown in arrangements 2 and 3 allow some adjustment, while the turnbuckle shown in arrangement 1 shows a common design for situations requiring large hanger adjustability.

The rod assembly may be attached to the building structure in various ways; some of these are illustrated in Fig. 6.4. The attachments shown are

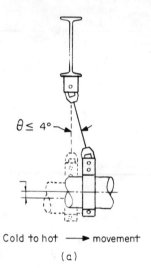

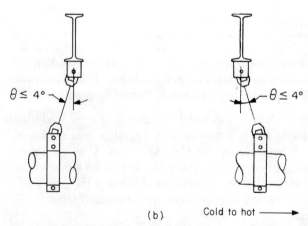

Figure 6.5 Offset limitations for rod hangers: (*a*) pipe with movement in axial direction; (*b*) cold pipe positioning—(1) offset, cold; (2) operating position, hot.

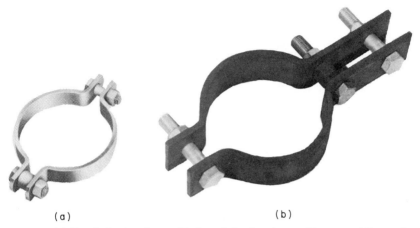

Figure 6.6 (a) Two-bolt pipe clamp; (b) three-bolt pipe clamp. *(Courtesy of Corner & Lada Co., Inc.)*

widely used in both commercial and nuclear design applications, with the type chosen according to point of attachment, desired adjustability, and amount of thermal growth. After the type of attachment and its orientation are determined, the size of attachment must also be verified by consulting the load ratings.

The recommended pipe attachments for weight supports on horizontal pipe runs are clamps or clevis hangers. Pipe clamps are used with rods as well as with spring hangers and rigid struts, which are discussed elsewhere in this chapter. Preengineered clamps are available for both commercial and nuclear piping.

The prime considerations for selecting a clamp are the pipe diameter, the design temperature, and the required load capacity. Nuclear-grade clamp specifications include material traceability and other quality assurance requirements. Clamps may be either two-bolt or three-bolt clamps (as shown in Fig. 6.6). Two-bolt clamps are usually used with 1 in (25.4 mm) of piping insulation or less; the three-bolt clamp permits the loading pin to protrude from thicker insulation.

The supplier of pipe support hardware usually supplies the load capacity data sheets (when required) and maintains the supporting data in their files for client reference.

Published design data indicates a load capacity based on a maximum temperature value, such as 650°F (343°C). Usually a manufacturer can supply a factor to correct the published value to reflect material allowable stresses at higher or lower temperatures.

When standard hardware does not fulfill the support design requirement, a nonstandard clamp can be used. Nonstandard clamps may be constructed of heavier stock for increased load capacity, may be smaller to fit into reduced-space envelopes, or may be of different materials in order to maintain compatibility with the piping material.

When the pipe is running vertically, normal clamps cannot be used for support in the vertical direction. In this case a riser clamp (see Fig. 6.7) would be used. Riser clamps support the pipe in the direction parallel to the axis of the pipe. On horizontal pipe runs, riser clamps may be used in axial supports. Since friction forces between the clamp and the pipe cannot be expected to exceed the support load, riser clamps must always be used in combination with shear lugs.

The load capacity data sheets for riser clamps indicate the total load applied to both arms, although most piping codes require that the arms be sized such that each individually can withstand the total load. Ideally, the applied load will be equal on both sides on the clamp, eliminating eccentric loading on the pipe. When riser clamps are used, the lugs attached to the pipe must also be checked for load capacity and local stress requirements (see Chap. 8).

When no overhead steel is convenient, it may be more practical to support the pipe from below, with a sliding support. Sliding supports accept the applied load strictly through bearing and allow the pipe to slide, in order to accommodate horizontal thermal pipe movements. The support may consist of any of a variety of saddle types, a trunnion attachment, rollers, or simply the pipe resting directly on restraint steel. The type of sliding support is chosen based on the thermal pipe movement, the pipe insulation thickness, and the distance from the pipe to the support steel. Examples of sliding supports are shown in Fig. 6.8.

When a sliding support is used, the support should be designed for friction forces in addition to the calculated design loads. The friction force may be calculated as the product of the coefficient of friction (normally between 0.3 and 0.7 for steel sliding against steel) and the force applied normal to the plane of movement (in this case the bearing load). For very flexible supports, use of this friction force may lead to the calculation of excessive loads and deflections. A more realistic representation of friction effects could be achieved in this circumstance by imposing a displacement

Figure 6.7 Two-bolt riser clamp. *(Courtesy of Corner & Lada Co., Inc.)*

Pipe Support Hardware   161

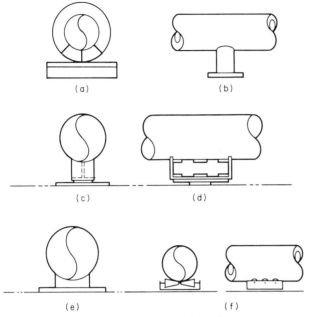

Figure 6.8 Types of sliding supports: (a) insulation protection saddle; (b) trunnion; (c), (d), (e) various types of saddles; (f) roller support.

equal to the pipe thermal movement at the support point onto the support structure, since the support deflection will be limited to the movement of the pipe.

Where it is desirable to reduce friction forces, lubricating plates may be placed between the pipe and the support. When lubricating plates are used, care should be taken to ensure that these materials are compatible with the expected environment. For example, Teflon, commonly used as a slide plate material, is known to degrade when exposed to high radiation and so must be replaced periodically.

6.2.2 Variable-spring supports

As noted previously, rod hangers and sliding supports may be used in locations where vertical thermal movements are minimal. However, where thermal pipe movements are large, the result of using a rigid support may be either the pipe lifting off of the support (and, therefore, loss of weight support) or thermal lockup at the support, with accompanying expansion overstress in the pipe. In these cases, it is necessary to support the piping system with spring hangers, which provide an upward force while still permitting the pipe to move.

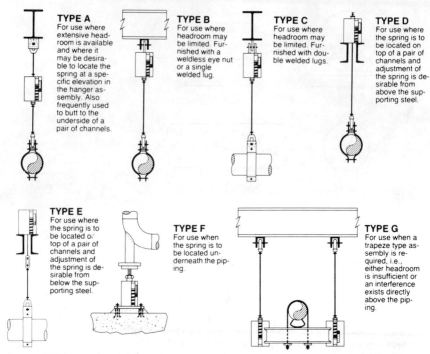

Figure 6.9 Types of variable-spring hangers. *(Courtesy of Corner & Lada Co., Inc.)*

Variable-spring hangers (see Figs. 6.9 and 6.10) are so called because they provide variable supporting forces as the pipe moves vertically. This is due to the elongation and contraction of the spring within the can assembly. The spring is initially precompressed prior to installation on the system; upward motion of the pipe causes spring extension and therefore reduces the spring force. Downward motion increases the spring compression, consequently increasing the resisting force.

The goal in pipe design applications is to provide a spring setting which will reach the correct weight load subsequent to the growth of the pipe to its normal operating position. Simultaneously, it is desired that the difference between the nonoperating "cold load" and the operating "hot load" not be too excessive, to prevent significant system imbalance.

The installation load, or cold load, of a variable spring may be calculated by the following formula:

$$\text{Cold load} = \text{hot load} + K\Delta \tag{6.1}$$

where cold load = installation setting lb (N)

hot load = weight load at support, determined from weight-balancing calculation plus weight of support hardware carried by spring, lb (N)

K = spring constant of variable spring, lb/in (N/mm)

Δ = expected vertical thermal movement of pipe at support point, from nonoperating to operating condition, where upward movement is considered positive, in (mm)

From this equation, it is evident that a spring at a pipe location which grows upward will have a cold load higher than the hot load, while for a spring at a location which moves downward the cold load will be lower. Remember that the direction of fluid flow in the piping system has noth-

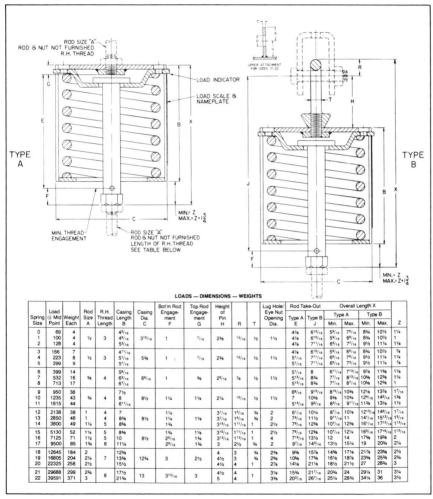

Figure 6.10 Standard spring types A and B—typical dimensions. *(Courtesy of Corner & Lada Co., Inc.)*

ing to do with the direction of the thermal expansion, which is rather a function of system geometry and restraint configuration.

As mentioned previously, it is desirable to limit the variability of the spring force between the hot and cold conditions. The variability is calculated as

$$V = \frac{K\Delta}{\text{hot load}} \tag{6.2}$$

Accepted practice in U.S. industry usually limits this variability to 25 percent, although on critical piping this may be limited to as low as 10 percent.

Since the hot load and thermal movement are dictated by the piping system configuration, the variability of the spring can be controlled only by varying the spring rate. Most manufacturers provide preengineered springs with three different spring rates per load size, recommended for short range [0 to 0.5-in (12.7-mm) movement], midrange [0.5 in (12.7 mm) to 1 in (25.4 mm)], and long range [1 in (25.4 mm) to 2 in (50.8 mm)] thermal movement, with spring rate ratios of 4:2:1, respectively.

Variable springs are selected by using a manufacturer's selection chart similar to that shown in Table 6.2. The procedure for selecting a variable-spring support is as described on Table 6.2.

6.2.3 Constant-spring supports

When thermal movements are too large [typically over 2 in (50.8 mm)] to permit the use of a variable spring, the engineer may elect to use a constant-spring support. Constant springs are also used when pipe stress concerns are critical or at locations near equipment at which very low nozzle loads must be maintained. These supports provide a virtually uniform supporting force throughout the travel range of the pipe. One method of achieving this constant load is through the use of a pulley and weight or a lever and weight system; however, these types of supports have disadvantages of additional weight requiring support and general lack of compactness.

Spring-loaded constant supports, as shown in Fig. 6.11, have most of the advantages and few of the drawbacks of the counterweight systems. Through the application of the principles of mechanical advantage, this type of spring support provides a nearly flat load response throughout its travel range, with variability deviation or load variation typically of the order of 5 percent.

The geometric design of the constant-support hanger assures perfectly constant support through the entire deflection of the pipe load. This counterbalancing of the load and spring moments about the main pivot is obtained by the use of carefully designed compression-type load springs, lever, and spring tension rods.

As the lever moves from the high to the low position, the load spring is compressed and the resulting increasing force acting on the decreasing spring moment arm creates a turning moment about the main pivot, which is exactly equal and opposite to the turning moment of the load and load moment arm.

As the lever moves from the low to the high position, the load spring is increasing in length and the resulting decreasing force acting on the increasing spring moment arm creates a turning moment about the main pivot which is exactly equal and opposite to the turning moment of the load and load moment arm.

Typical constant-spring support applications are illustrated in Fig. 6.12. The particular design utilized will usually depend on the space constraints which dictate whether the support is to be installed above, between, or below the building steel.

Constant-spring supports are normally available in various sizes to accommodate the required load. The load is calculated as is the hot load of the variable-spring hanger—the load from the weight-balancing calculation plus the weight of the support hardware. The constant support size is selected based on pipe thermal movement and the weight load, by using a selection chart such as the one shown in Table 6.3. The supporting load of the spring may be either preset in the factory or set during installation in the field. The total travel of the spring should be conservatively set at the calculated thermal movement of the pipe plus 20 percent, or 1 in (25.4 mm), whichever is greater.

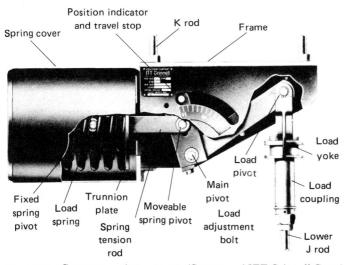

Figure 6.11 Constant-spring support. *(Courtesy of ITT Grinnell Corp.)*

TABLE 6.2 Selection of Variable Springs

The key criterion in selecting the size and series of a variable spring is a factor known as variability. This is a measurement of the percentage change in supporting force between the hot and cold positions of a spring and is calculated from the formula

$$\text{Variability} = \frac{\text{hot load} - \text{cold load}}{\text{hot load}}$$

The cold load is calculated by adding (for up movement) or subtracting (for down movement) the product of spring rate times movement to or from the hot load. If an allowable variability is not specified, good practice would be to use 25% for noncritical piping and 10% for critical piping.

Working range deflection, in.				Hanger size										
Long-range	Mid-range	Short-range		0	1	2	3	4	5	6	7	8	9	10
			Reserve range	43	64	82	106	142	190	252	336	450	600	780
				44	66	84	109	146	196	260	347	465	620	806
				46	68	87	113	151	202	269	358	480	640	832
				47	70	90	116	156	208	277	370	495	660	858
				49	72	92	120	160	215	286	381	510	680	884
0.0	0.0	0.0		50	74	95	123	165	221	294	392	525	700	910
0.2	0.1			52	76	98	127	170	227	302	403	540	720	926
0.4	0.2	0.1		53	78	100	130	174	234	311	414	555	740	962
0.6	0.3			55	80	103	134	179	240	319	426	570	760	988
0.8	0.4	0.2		56	82	106	137	184	246	328	437	585	780	1,014
1.0	0.5			57	84	108	140	188	252	336	448	600	800	1,040
1.2	0.6	0.3		59	87	111	144	193	259	344	459	615	820	1,066
1.4	0.7			60	89	114	147	198	265	353	470	630	840	1,092
1.6	0.8	0.4	Working range	62	91	117	151	203	271	361	482	645	860	1,118
1.8	0.9			63	93	119	154	207	278	370	493	660	880	1,144
2.0	1.0	0.5		65	95	122	158	212	284	378	504	675	900	1,170
2.2	1.1			66	97	125	161	217	290	386	515	690	920	1,196
2.4	1.2	0.6		68	99	127	165	221	297	395	526	705	940	1,222
2.6	1.3			69	101	130	168	226	303	403	538	720	960	1,248
2.8	1.4	0.7		71	103	133	172	231	309	412	549	735	980	1,274
3.0	1.5			72	105	135	175	235	315	420	560	750	1,000	1,300
3.2	1.6	0.8		74	108	138	179	240	322	428	571	765	1,020	1,326
3.4	1.7			75	110	141	182	245	328	437	582	780	1,040	1,352
3.6	1.8	0.9		77	112	144	186	250	334	445	594	795	1,060	1,378
3.8	1.9			78	114	146	189	254	341	454	605	810	1,080	1,404

To size a spring:
1. Calculate the maximum allowable spring rate from the formula:

$$\text{Spring rate} = \frac{\text{variability} \times \text{hot load}}{\text{movement}}$$

2. Determine the size by finding the hot load in the chart.
3. Stay in that size column and choose the series with a spring rate equal to or less than the value calculated above.
4. Calculate the cold load and check that both hot and cold loads fall within the working range.
5. If this condition is not met, move to an adjacent size and rework.

If load, movement, variability, or available space prohibits the use of a variable spring, the use of a constant support should be considered.

Hanger size												Total spring deflection, in.		
11	12	13	14	15	16	17	18	19	20	21	22	Short-range	Mid-range	Long-range
1,020	1,350	1,800	2,400	3,240	4,500	6,000	7,990	10,610	14,100	18,750	25,006	0.0	0.0	0.0
1,054	1,395	1,860	2,480	3,348	4,650	6,200	8,256	10,964	14,570	19,375	25,839		0.1	0.2
1,088	1,440	1,920	2,560	3,456	4,800	6,400	8,522	11,318	15,040	20,000	26,673	0.1	0.2	0.4
1,122	1,485	1,980	2,640	3,564	4,950	6,600	8,788	11,672	15,510	20,625	27,506		0.3	0.6
1,156	1,530	2,040	2,720	3,672	5,100	6,800	9,054	12,026	15,980	21,250	28,340	0.2	0.4	0.8
1,190	1,575	2,100	2,800	3,780	5,250	7,000	9,320	12,380	16,450	21,875	29,173		0.5	1.0
1,224	1,620	2,160	2,880	3,888	5,400	7,200	9,586	12,734	16,920	22,500	30,007	0.3	0.6	1.2
1,258	1,665	2,220	2,960	3,996	5,550	7,400	9,852	13,088	17,390	23,125	30,840		0.7	1.4
1,292	1,710	2,280	3,040	4,104	5,700	7,600	10,118	13,442	17,860	23,750	31,674	0.4	0.8	1.6
1,326	1,755	2,340	3,120	4,212	5,850	7,800	10,384	13,796	18,330	24,375	32,507		0.9	1.8
1,360	1,800	2,400	3,200	4,320	6,000	8,000	10,650	14,150	18,800	25,000	33,340	0.5	1.0	2.0
1,394	1,845	2,460	3,280	4,428	6,150	8,200	10,916	14,504	19,270	25,625	34,174		1.1	2.2
1,428	1,890	2,520	3,360	4,536	6,300	8,400	11,182	14,858	19,740	26,250	35,007	0.6	1.2	2.4
1,462	1,935	2,580	3,440	4,644	6,450	8,600	11,448	15,212	20,210	26,875	35,841		1.3	2.6
1,496	1,980	2,640	3,520	4,752	6,600	8,800	11,714	15,566	20,680	27,500	36,674	0.7	1.4	2.8
1,530	2,025	2,700	3,600	4,860	6,750	9,000	11,980	15,921	21,250	28,125	37,508		1.5	3.0
1,564	2,070	2,760	3,680	4,968	6,900	9,200	12,246	16,274	21,620	28,750	38,341	0.8	1.6	3.2
1,598	2,115	2,820	3,760	5,076	7,050	9,400	12,512	16,628	22,090	29,375	39,175		1.7	3.4
1,632	2,160	2,880	3,840	5,184	7,200	9,600	12,778	16,982	22,560	30,000	40,008	0.9	1.8	3.6
1,666	2,205	2,940	3,920	5,292	7,350	9,800	13,044	17,336	23,030	30,625	40,842		1.9	3.8
1,700	2,250	3,000	4,000	5,400	7,500	10,000	13,310	17,690	23,500	31,250	41,675	1.0	2.0	4.0
1,734	2,295	3,060	4,080	5,508	7,650	10,200	13,576	18.044	23,970	31,875	42,509		2.1	4.2
1,768	2,340	3,120	4,160	5,616	7,800	10,400	13,842	18,398	24,440	32,500	43,342	1.1	2.2	4.4
1,802	2,385	3,180	4,240	5,724	7,950	10,600	14,108	18.752	24,910	33,125	44,176		2.3	4.6
1,836	2,430	3,240	4,320	5,832	8,100	10,800	14,374	19,106	25,380	33,750	45,009	1.2	2.4	4.8

TABLE 6.2 Selection of Variable Springs (Continued)

Working range deflection, in.				Hanger size										
Long-range	Mid-range	Short-range		0	1	2	3	4	5	6	7	8	9	10
4.0	2.0	1.0		80	116	149	193	259	347	462	616	825	1,100	1,430
4.2	2.1		Working range	81	118	152	196	264	353	470	627	840	1,120	1,456
4.4	2.2	1.1	Working range	83	120	154	200	268	360	479	638	855	1,140	1,482
4.6	2.3		Working range	84	122	157	203	273	366	487	650	870	1,160	1,508
4.8	2.4	1.2	Working range	86	124	160	207	278	372	496	661	885	1,180	1,534
5.0	2.5		Working range	88	126	162	210	282	378	504	672	900	1,200	1,560
				89	129	165	214	287	385	512	683	915	1,220	1,586
			Reserve range	90	131	168	217	292	391	521	694	930	1,240	1,612
			Reserve range	92	133	171	221	297	397	529	706	945	1,260	1,638
			Reserve range	93	135	173	224	301	404	538	717	960	1,280	1,664
				95	137	176	228	306	410	546	728	975	1,300	1,690
			Spring Rate, Pounds Per Inch of Deflection											
				30	42	54	70	94	126	168	224	300	400	520
				15	21	27	35	47	63	84	112	150	200	260
				7	10	13	17	23	31	42	56	75	100	130

SOURCE: Courtesy of Corner & Lada Co., Inc.

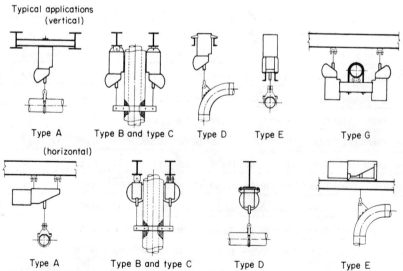

Figure 6.12 Typical constant-spring applications. (*Courtesy of ITT Grinnell Corp.*)

				Hanger size							Total spring deflection, in.		
11	12	13	14	15	16	17	18	19	20	21	22	Short-range	Mid-range	Long-range
1,870	2,475	3,300	4,400	5,940	8,250	11,000	14,640	19,460	25,850	34,375	45,843		2.5	5.0
1,904	2,250	3,360	4,480	6,048	8,400	11,200	14,906	19,814	26,320	35,000	46,676	1.3	2.6	5.2
1,938	2,565	3,420	4,560	6,156	8,550	11,400	15,172	20,168	26,790	35,625	47,510		2.7	5.4
1,972	2,610	3,480	4,640	6,264	8,700	11,600	15,438	20,522	27,260	36,250	48,343	1.4	2.8	5.6
2,006	2,655	3,540	4,720	6,372	8,850	11,800	15,704	20,876	27,730	36,875	49,177		2.9	5.8
2,040	2,700	3,600	4,800	6,480	9,000	12,000	15,970	21,230	28,200	37,500	50,010	1.5	3.0	6.0
2,074	2,745	3,660	4,880	6,588	9,150	12,200	16,236	21,584	28,670	38,125	50,844		3.1	6.2
2,108	2,790	3,720	4,960	6,696	9,300	12,400	16,502	21,938	29,140	38,750	51,677	1.6	3.2	6.4
2,142	2,835	3,780	5,040	6,804	9,450	12,600	16,768	22,292	29,610	39,375	52,511		3.3	6.6
2,176	2,880	3,840	5,120	6,912	9,600	12,800	17,034	22,646	30,080	40,000	53,344	1.7	3.4	6.8
2,210	2,925	3,900	5,200	7,020	9,750	13,000	17,300	23,000	30,550	40,625	54,178		3.5	7.0
				Spring Rate, Pounds Per Inch of Deflection										
680	900	1,200	1,600	2,160	3,000	4,000	5,320	7,080	9,400	12,500	16,670			
340	450	600	800	1,080	1,500	2,000	2,660	3,540	4,700	6,250	8,335			
170	225	300	400	540	750	1,000	1,330	1,770	2,350	3,125	4,167			

6.3 Rigid Restraints

Rigid restraints are used when it is necessary to resist weight, thermal, and occasional loads at a single location or where thermal movements are small enough to permit use of a rigid restraint. Normally it is more economical to use a rigid restraint than one which permits movement, such as a spring or a snubber assembly.

Rigid restraints may be provided by using either preengineered hardware (rigid struts) or rigid frames built up of structural steel. Wide-flange beams, back-to-back channels, or tubes are commonly used structural steel sections in rigid frames. Angle sections are sometimes used, though the unsymmetric nature of their cross section requires added design procedures. The decision of whether to use a strut or to restrain the pipe with a steel frame is normally dictated by considerations such as space constraints, number of directions in which the pipe must be simultaneously restrained, lead time for hardware delivery, etc. Where possible, use of load-rated preengineered hardware is preferred since it eliminates the costly engineering time required to design a steel structure.

TABLE 6.3 Constant-Support Selection Chart

Hanger size	\	\	\	\	\	\	\	Load in pounds for total travel in inches	\	\	\	\	\	\	\
	1½	2	2½	3	3½	4	4½	5	5½	6	6½	7	7½	8	8½
1	144	108	86	72	62	54	48	43	39	36	33	31	29	27	
	173	130	104	87	74	65	58	52	47	43	40	37	35	33	
2	204	153	122	102	87	77	68	61	56	51	47	44	41	36	
3	233	175	140	117	100	88	78	70	64	58	54	50	47	44	
4	280	210	168	140	120	105	93	84	76	70	65	60	56	53	
5	327	245	196	163	140	123	109	96	89	82	75	70	65	61	
6	373	280	224	187	160	140	124	112	102	93	86	80	75	70	
7	451	338	270	225	193	169	150	135	123	113	104	97	90	85	
8	527	395	316	263	226	196	176	158	144	132	122	113	105	99	
9	600	450	360	300	257	225	200	180	164	150	138	129	120	113	
10	727	545	436	363	311	273	242	218	198	182	168	156	145	136	
11	851	638	510	425	365	319	284	255	232	213	196	182	170	160	
12	977	733	586	489	419	367	326	293	267	244	226	209	195	183	
13	1,177	883	706	589	505	442	392	353	321	294	272	252	235	221	
14	1,373	1,030	824	687	589	515	458	412	375	343	317	294	275	258	
15	1,573	1,180	944	787	674	590	524	472	429	393	363	337	315	295	

16	1,893	1,420	1,136	947	811	710	631	568	516	473	327	406	379	355		
17	2,217	1,663	1,330	1,109	950	832	739	665	605	554	512	475	443	416		
18	2,540	1,905	1,524	1,270	1,089	953	847	762	693	635	586	544	508	476	448	
19		2,025	1,620	1,350	1,157	1,013	900	810	736	675	623	579	540	506	476	
20		2,145	1,716	1,430	1,226	1,073	953	858	780	715	660	613	572	536	505	
21		2,335	1,868	1,557	1,334	1,168	1,038	934	849	778	718	667	623	584	549	
22		2,525	2,020	1,683	1,442	1,263	1,122	1,010	918	842	777	721	673	631	594	
23		2,710	2,168	1,807	1,549	1,355	1,204	1,064	985	903	834	775	723	678	638	
24		2,910	2,328	1,940	1,663	1,455	1,293	1,164	1,056	970	958	831	776	728	685	
25		3,110	2,488	2,073	1,777	1,555	1,382	1,244	1,131	1,037	957	889	829	778	732	
26		3,310	2,648	2,207	1,891	1,655	1,471	1,324	1,204	1,103	1,018	946	883	828	779	
27		3,630	2,904	2,420	2,074	1,815	1,613	1,452	1,320	1,210	1,117	1,037	968	908	854	
28		3,950	3,160	2,633	2,257	1,975	1,756	1,580	1,436	1,317	1,215	1,129	1,053	968	929	
29		4,270	3,416	2,847	2,440	2,135	1,896	1,708	1,553	1,423	1,314	1,220	1,139	1,068	1,005	
30		4,535	3,628	3,023	2,591	2,268	2,016	1,814	1,649	1,512	1,395	1,296	1,209	1,134	1,067	
31		4,795	3,836	3,197	2,740	2,398	2,131	1,918	1,744	1,598	1,475	1,370	1,279	1,199	1,128	
32		5,060	4,048	3,373	2,891	2,530	2,249	2,024	1,840	1,687	1,557	1,446	1,349	1,265	1,191	
33		5,295	4,236	3,530	3,026	2,648	2,353	2,118	1,765	1,965	1,629	1,513	1,412	1,324	1,245	

TABLE 6.3 Constant-Support Selection Chart (*Continued*)

Hanger size	\multicolumn{16}{c}{Load in pounds for total travel in inches}															
	$1\frac{1}{2}$	2	$2\frac{1}{2}$	3	$3\frac{1}{2}$	4	$4\frac{1}{2}$	5	$5\frac{1}{2}$	6	$6\frac{1}{2}$	7	$7\frac{1}{2}$	8	$8\frac{1}{2}$
34	5,625		4,420	3,683	3,157	2,763	2,456	2,210	2,009	1,842	1,700	1,579	1,473	1,381	1,300
35			4,696	3,913	3,354	2,935	2,609	2,348	2,135	1,957	1,806	1,677	1,565	1,468	1,381
36			4,968	4,140	3,549	3,105	2,760	2,484	2,258	2,070	1,911	1,774	1,656	1,553	1,416
37			5,240	4,367	3,743	3,275	2,911	2,620	2,382	2,183	2,015	1,871	1,747	1,638	1,541
38			5,616	4,680	4,011	3,510	3,120	2,808	2,553	2,340	2,160	2,006	1,872	1,755	1,652
39			5,988	4,990	4,277	3,743	3,327	2,994	2,722	2,495	2,303	2,139	1,996	1,871	1,761
40			6,360	5,300	4,543	3,975	3,533	3,180	2,891	2,650	2,446	2,271	2,120	1,988	1,871
41			6,976	5,813	4,983	4,360	3,876	3,488	3,171	2,907	2,683	2,491	2,325	2,180	2,052
42			7,588	6,323	5,420	4,743	4,216	3,794	3,449	3,162	2,919	2,710	2,529	2,371	2,232
43			8,200	6,833	5,857	5,125	4,556	4,100	3,727	3,417	3,154	2,929	2,733	2,583	2,412
44			8,724	7,270	6,231	5,453	4,847	4,362	3,965	3,635	3,355	3,116	2,908	2,726	2,566
45			9,284	7,737	6,831	5,803	5,158	4,642	4,220	3,868	3,571	3,316	3,095	2,901	2,731
46			9,760	8,133	6,971	6,100	5,422	4,880	4,436	4,067	3,754	3,486	3,253	3,050	2,871
47			10,376	8,647	7,411	6,485	5,764	5,188	4,716	4,323	3,991	3,706	3,459	3,243	3,052
48			10,988	9,157	7,848	6,868	6,104	5,494	4,995	4,578	4,226	3,924	3,663	3,434	3,232

49		11,600	9,667	8,286	7,250	6,444	5,800	5,273	4,833	4,462	4,143	3,867	3,625	3,412
50			10,367	8,886	7,775	6,911	6,220	5,855	5,183	4,785	4,443	4,147	3,886	3,659
51			11,067	9,486	8,300	7,378	6,640	6,036	5,533	5,108	4,743	4,427	4,150	3,906
52			11,847	10,154	8,885	7,896	7,108	6,462	5,923	5,473	5,077	4,739	4,443	4,181
53			12,623	10,820	9,468	8,415	7,574	6,886	6,311	5,826	5,410	5,049	4,734	4,455
54			13,400	11,486	10,050	8,933	8,040	7,309	6,700	6,185	5,743	5,360	5,025	4,730
55			14,713	12,611	11,035	9,809	8,828	8,026	7,356	6,791	6,306	5,885	5,518	5,193
56			16,023	13,734	12,018	10,682	9,614	8,740	8,011	7,396	6,867	6,409	6,009	5,855
57			17,333	14,857	13,000	11,555	10,400	9,455	8,666	8,000	7,429	6,933	6,500	6,118
58			18,423	15,791	13,818	12,282	11,054	10,049	9,211	8,503	7,896	7,369	6,809	6,503
59			19,510	16,723	14,633	13,007	11,706	10,642	9,755	9,005	8,362	7,804	7,316	6,886
60			20,600	17,657	15,450	13,733	12,360	11,238	10,300	9,508	8,829	8,240	7,725	7,271
61			21,890	18,763	16,418	14,593	13,134	11,940	10,945	10,103	9,382	8,756	8,209	7,726
62			23,176	19,665	17,383	15,451	13,906	12,642	11,588	10,697	9,933	9,270	8,691	8,180
63			24,463	20,968	18,348	16,309	14,678	13,344	12,231	11,291	10,484	9,785	9,174	8,634

SOURCE: *Courtesy of ITT Grinnell Corp.*

The strut and clamp assembly shown in Fig. 6.13 is sized for load capacity as well as stiffness criteria, if necessary. Initially, the strut assembly is selected based on the applied load. If the strut does not meet required stiffness criteria, a larger strut size should be selected. The typical dimensions shown in Fig. 6.13 are used to calculate the strut lengths. The designer must ensure that adequate space is available when a strut is used. The strut is made up of two end "paddles," with bushings for pins, welded to an intermediate pipe of varying length. In most designs, one paddle is threaded to provide some field adjustability. The failure mode which limits the capacity of the strut is normally buckling of the pipe. For this reason, the vendor's load rating is usually valid only up to a certain specified length, the given pin-to-pin distance (L in Fig. 6.13). At longer lengths, the capacity of the strut will decrease as the slenderness ratio increases, to account for the reduction in buckling load capacity. Strut capacities at longer lengths are normally available from the manufacturer.

The strut rating is based on a tolerance of offset from the line of action of the applied force, normally stated as ±6° from the installation reference axis. Therefore, in certain situations, such as when tension is the controlling design load, when the length of the strut is significantly below the maximum length for the load rating, or when it is known that there will

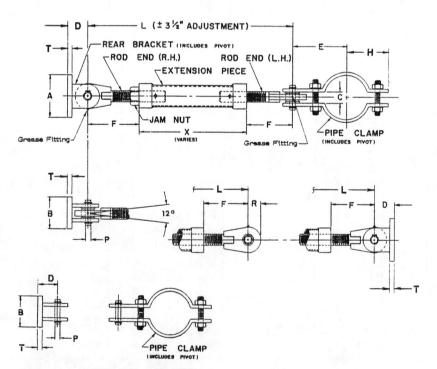

Figure 6.13 Rigid strut assembly. *(Courtesy of Corner & Lada Co., Inc.)*

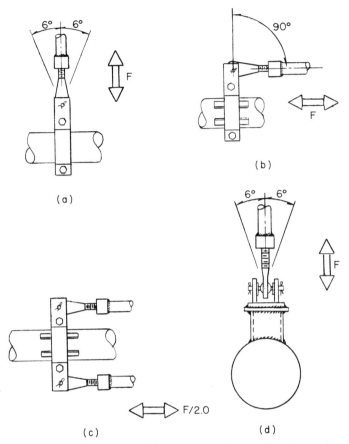

Figure 6.14 Rigid strut assemblies—load application directions: (*a*) radial load clamp; (*b*) eccentric load clamp with shear lugs; (*c*) riser clamp with shear lugs; (*d*) strut bracket welded to trunnion.

be no strut offset, the engineer may choose to upgrade the capacity of the strut (for the specific application) through a review of its component parts.

Struts may be attached to the pipe through either clamps or welded attachments. Clamps are available in three basic types, with stock and pin size varying according to load capacity. The first type of clamp is for loads applied radially ($\pm 6°$) to the pipe, as shown in Fig. 6.14*a*. The second type, with a wider stock and an eccentric pin position, is capable of resisting the moment induced by an eccentric loading. With this type of clamp (see Fig. 6.14*b*) shear lugs are required to prevent the clamp from sliding along the pipe. When an axial restraint is required and an eccentric loading is undesirable, a riser clamp arrangement should be used (named such because these types of clamps are commonly used to support risers). The clamps accept two struts (one on each side of the pipe), thus transferring a zero net bending moment on the pipe, as shown in Fig. 6.14c. In lieu of a clamp, the strut bracket may be welded directly to a welded pipe attachment, such as the trunnion in Fig. 6.14d. This configuration permits orientation of the strut anywhere within a 180° range relative to the pipe axis and within ±6° off the axis.

Struts may be attached directly to building steel, baseplates on concrete structures, or built-up support structures, as shown in Fig. 6.15. Where support is required in the direction of two perpendicular axes, two struts may be used, oriented anywhere in the plane of those axes. The loads on the individual struts can be resolved through static equilibrium, as follows:

$$\Sigma F_1 = 0: \quad \frac{6}{(6^2 + 5.5^2)^{1/2}} F_A - F_1 = 0$$

$$\Sigma M = 0: \quad 5.5 F_1 - 6 F_2 + 6 F_B = 0$$

By substitution,

$$F_A = 1.3566 F_1 \quad \text{and} \quad F_B = F_2 - 0.9167 F_1$$

In order to maintain an economical design, it is best to use an included angle between the struts of 30 to 120°.

Hardware other than struts, when it is used in conjunction with structural steel, may provide rigid restraint. One of the most common types of

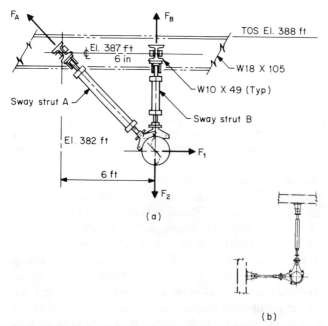

Figure 6.15 Application of rigid strut and structural steel combination.

Pipe Support Hardware   177

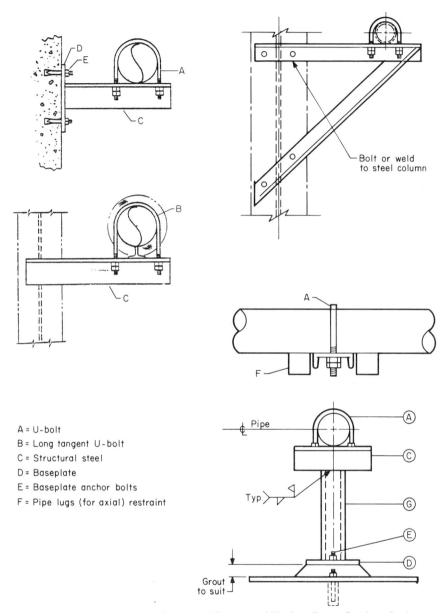

A = U-bolt
B = Long tangent U-bolt
C = Structural steel
D = Baseplate
E = Baseplate anchor bolts
F = Pipe lugs (for axial) restraint

Figure 6.16 Typical U-bolt applications. *(Courtesy of Nuclear Power Services, Inc.)*

restraining devices is the U-bolt, a rod bent in the shape of a U, as shown in Fig. 6.16.

The U-bolt provides restraint in two directions, although the shear capacity of the U-bolt is usually well below the tensile capacity. For this reason, U-bolts are usually restricted to applications where the lateral load is minimal, such as on piping not subject to seismic loading, where thermal movements are expected to be small, or on restraints using trapeze assemblies which are free to move laterally themselves. Tensile and shear capacities are normally stated in the vendor's catalog. When evaluating the U-bolt for bidirectional loadings, the engineer should use a linear interaction formula such as

$$\frac{T_{\text{applied}}}{T_{\text{allowable}}} + \frac{S_{\text{applied}}}{S_{\text{allowable}}} \leq 1.0 \tag{6.3}$$

where T_{applied} and $T_{\text{allowable}}$ are the applied and allowable (provided by vendor) tensile forces on the U-bolt, respectively, and S_{applied} and $S_{\text{allowable}}$ are the applied and allowable (provided by vendor) shear (lateral) forces on the U-bolt, respectively.

In cases where the lateral load exceeds the capacity of the U-bolt, the engineer may choose to use a pipe strap, as shown in Fig. 6.17. The pipe strap is also in the shape of a U; however, it is made of bar stock which, when it is welded to the support steel, provides greater resistance to lateral loads. When support is desired in one direction only, a one-directional pipe strap may be used very effectively.

Another common method of providing rigid restraint to a pipe is through direct contact with the support steel. This may be done by sup-

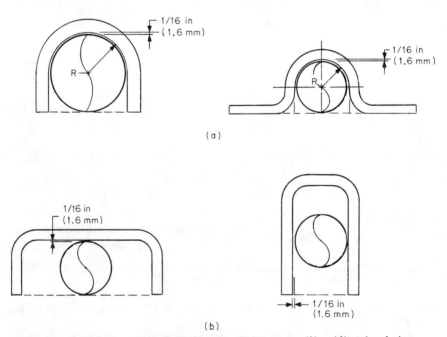

Figure 6.17 Typical pipe straps: (*a*) bidirectional pipe straps; (*b*) unidirectional pipe straps. *(Courtesy of Corner & Lada Co., Inc.)*

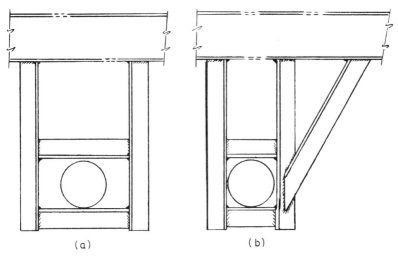

Figure 6.18 Use of structural steel for support: (*a*) vertical restraint; (*b*) vertical and lateral restraint.

porting the pipe directly on steel, boxing it in, or by transmitting the force to the steel through welded attachments. Figure 6.18 shows two examples of support provided through the use of structural steel.

Lateral loads may be resisted by boxing in the pipe directly or by transmitting the load through pipe attachments or bearing lugs. The support steel, when it is in contact with the pipe, must be sized not only to adequately resist the load, but also to provide sufficient linear contact with the pipe to satisfy local pipe wall stress criteria. Likewise, bearing lugs must be sized with both the load and the local stress considerations in mind. When bearing lugs are designed, it is often necessary to consider bending moments on the pipe due to friction between the lug and the steel when the pipe moves thermally.

Axial-direction restraint of the pipe cannot readily be provided with only support steel. The load must usually be transmitted to the steel through shear lugs welded to the pipe. These lugs, similar in function to those found with eccentrically loaded clamps, must be sized according to both strength and local pipe wall stress considerations. The allowable pipe stress for lug design may usually be determined by deducting the pipe stresses caused by the piping design conditions from the total allowable stress of the pipe material.

Welded attachments often facilitate the support of multiple degrees of freedom with a single support, and certain support functions (notably axial or torsional supports) are nearly impossible to provide without welded attachments to the pipe. However, welded attachments have drawbacks in that (1) high local stresses may be induced in the pipe wall

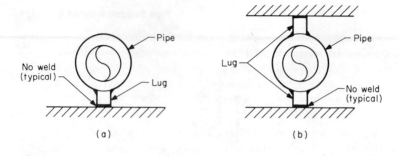
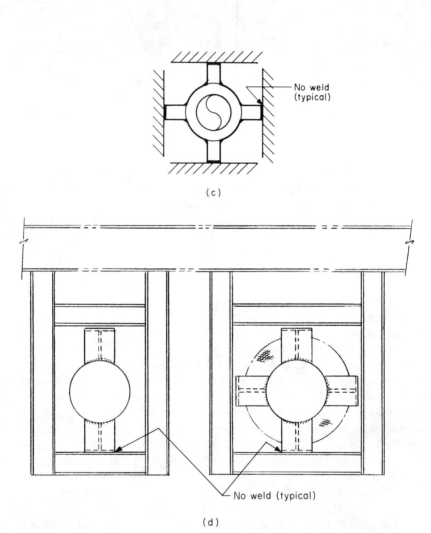

Figure 6.19 Bearing lugs: (*a*) single lug; (*b*) pair of lugs; (*c*) four lugs; (*d*) bearing lugs used with structural steel.

due to the attachment, (2) the welds require frequent inspection as they become part of the pressure boundary of the pipe, and (3) welded attachments may not be installed during system operation (such as on backfit jobs). For these reasons, welded attachments should be avoided where possible.

The most common types of welded attachments are bearing lugs, shear lugs, and trunnions. Bearing lugs are plates welded to the pipe, the purpose of which is to transmit the piping load to the restraint through bearing. These lugs are useful in resisting lateral loads. Shear lugs transmit loads to the support steel through shear and therefore are useful in designing axial restraints. Trunnions are pipe sections welded to the run pipe, to which pipe support hardware may be attached. Examples of supports using welded attachments are shown in Figs. 6.19 and 6.20.

When all degrees of freedom (three translation and three rotational) of

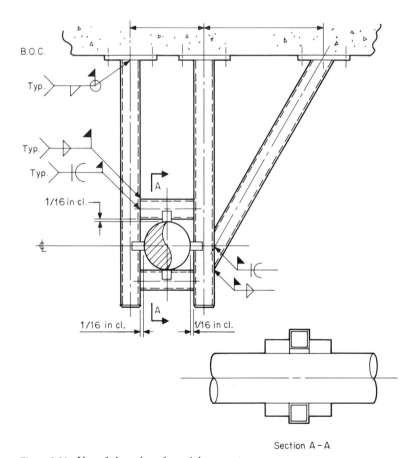

Figure 6.20 Use of shear lugs for axial support.

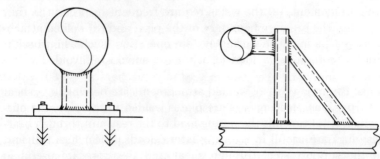

Figure 6.21 Typical anchor configurations.

the pipe are simultaneously supported at a single location, the resulting pipe support is called an *anchor*. An anchor provides a fixed reference point of constant position and rotation through which effects from the pipe on opposite sides cannot be transmitted. This makes the anchor a convenient terminal point for defining stress analysis problems. The anchor as shown in Fig. 6.21 can normally be built only with structural steel and welded pipe attachments as opposed to preengineered hardware.

6.4 Snubbers

Rigid restraints are usually necessary when the pipe is designed to survive horizontal loads, such as earthquakes or high winds, or other dynamic loads, such as fluid hammer. However, when the pipe has a high operating temperature, it is likely that at some locations a rigid restraint may not be installed without causing thermal lock-up and the accompanying elevated stress range levels in the pipe. In these cases, snubbers may be used in a way similar to the use of springs when thermal movement prevents the use of a rigid hanger.

Snubbers (see Figs. 6.22 and 6.23) resemble rigid struts, except that they contain a mechanism which permits movement in the presence of static or slowly applied loads, but which locks up during rapidly applied loads. Therefore the snubber will resist dynamic loads while permitting the natural and slower thermal growth of the pipe. Note that a snubber will not act as a weight support because of this design feature: when both weight and vertical dynamic restraint are required at a point of large vertical thermal movement, the engineer must provide both a spring and a snubber. The spring support will carry the weight load while the snubber will restrain the dynamic loads, with both supports permitting the required thermal movements.

Two types of snubbers are commonly available: hydraulic and mechanical. The hydraulic snubber is made up of a piston and a double-chamber reservoir filled with a viscous fluid. The stroke of a piston forces the fluid through a narrow passage between the two chambers. When the piston velocity, and therefore the fluid flow rate, reaches a threshold point, a valve closes the passage between the two chambers. From this point, the applied snubber force is resisted by the compressibility of the fluid. Mechanical snubbers operate on a purely mechanical basis, with no hydraulic fluid. They are designed to provide resistance when a threshold *acceleration* is reached; beneath that acceleration free movement of the pipe is permitted.

Hydraulic snubbers have been subject to loss of fluid due to seal design problems, degradation of hydraulic fluid in radioactive environments, and premature locking due to corrosion (or even painting, though this is not confined to hydraulic snubbers). Mechanical snubbers have also been known to lock up inadvertently, leading to additional piping stresses which could be detrimental to the system.

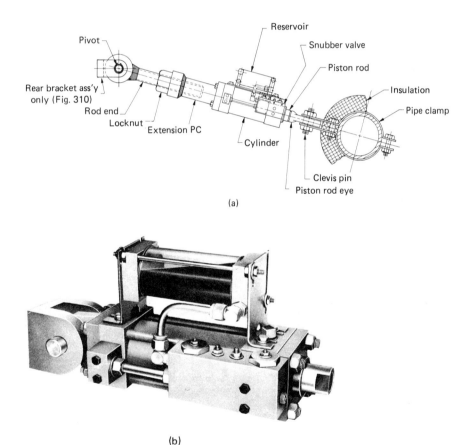

Figure 6.22 Hydraulic snubber: (*a*) schematic; (*b*) photograph. *(Courtesy of ITT Grinnell Corp.)*

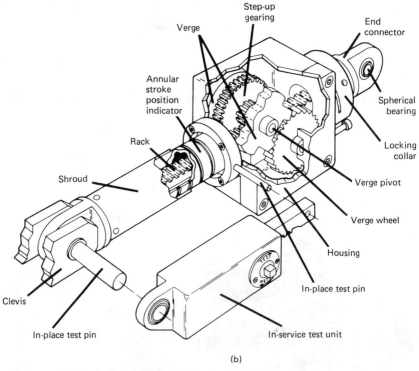

Figure 6.23 Mechanical snubber: (*a*) photograph; (*b*) schematic. *(Courtesy of Anchor/Darling Industries, Inc.)*

There are other considerations to bear in mind when hydraulic snubbers are used. When two hydraulic snubbers are used in tandem (see Fig. 6.24a), it cannot be guaranteed that both snubbers will lock simultaneously; therefore each snubber must be sized to take the full applied load, to prevent failure prior to locking of the second snubber. A second consideration is that, owing to the damping nature of the hydraulic snubber, the resisting force will not develop until a sufficient velocity develops. The mechanical snubber, however, should lock with the initial application of the disturbing acceleration. This will permit a greater displacement of the pipe during a shock loading when hydraulic snubbers are used than when mechanical snubbers are used.

Because of these problems, the nuclear power industry has, in the past few years, become concerned about the use of snubbers (both mechanical and hydraulic) for restraining piping systems. One area of concern is the consequence of an inoperable snubber increasing the probability of structural damage to piping in the event of dynamic loads. To assure that all snubbers located on systems required for the safe operation of the nuclear power plant are operable, intensive inspection and functional testing have been, and will continue to be, required by the NRC.

Recently the Task Group on Industry Practice of the PVRC Technical Committee on Piping Systems issued a draft position paper which states in part:

> Qualitatively, reliability as related to snubbers is simple: A system without any of these devices is more reliable than a system with them. Nonetheless, it must be recognized that some snubbers are needed so that piping systems located in confined spaces can be designed to be flexible enough to absorb thermal expansion loads and, at the same time, be rigid enough to withstand the dynamic loads imposed on them.

Recent emphasis on optimizing snubbers and restraints on both operating plants and plants under construction is a result of the high cost associated with snubbers, particularly due to costly installation, testing, and inspection requirements which are forced on operating nuclear power plants. Additionally, in-service inspection requirements for snubbers increase radiation exposure with little increase in plant safety, reliability, or economic output.

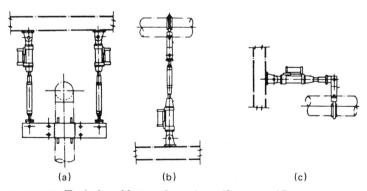

Figure 6.24 Typical snubber configurations. *(Courtesy of Bergen-Paterson Pipesupport Corp.)*

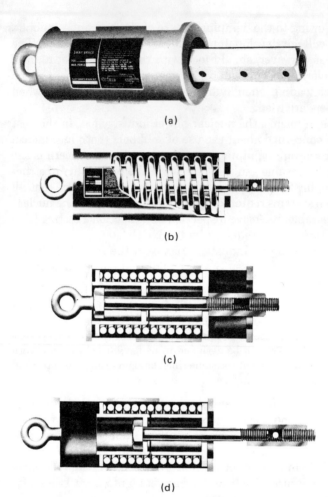

Figure 6.25 (*a*) Vibration control and sway brace; (*b*) cutaway section; (*c*) deflection of single spring occurs when thrust exceeds precompression; (*d*) tension causes deflection of single spring in opposite direction. *(Courtesy of Corner & Lada Co., Inc.)*

6.5 Sway Braces

Sway braces are used to limit the effects of pipe vibration and are ideally suited to those situations where precise control of impact forces is not required. These units (see Fig. 6.25) are little more than variable springs acting in the horizontal plane. When the sway brace is installed, the spring preload is adjusted to be zero when the pipe is in the operating position. As the pipe begins to displace during vibration, the sway brace will provide a restoring load equal to the spring constant of the brace times the pipe displacement. The effect of the sway brace on the piping system is to increase the K value in this equation:

$$M\ddot{X}(t) + C\dot{X}(t) + KX(t) = F(t) \tag{6.4}$$

This, in turn, will raise the natural frequencies of the vibratory modes and thus normally reduce the response of the pipe to dynamic loads and vibrations.

Sway braces, like variable springs, do add somewhat to the thermal expansion stresses in the pipe (although to a lesser degree than would rigid restraints). It is wise to minimize this effect by selecting sway brace locations which coincide with points of low thermal movements in the direction of restraint.

6.6 Baseplates

Aside from the hardware identified previously, other miscellaneous items are found in pipe supports. One of these is the baseplate anchor-bolt assembly.

Baseplates may either be of the anchor-bolted type (see Fig. 6.26a) or the embedded type (Fig. 6.26b). Baseplates with anchor bolts are normally used in cases where the building concrete has already been poured

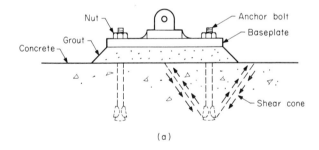

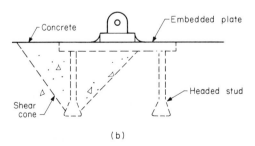

Figure 6.26 Baseplate types: (a) baseplate with anchor bolts; (b) embedded plate.

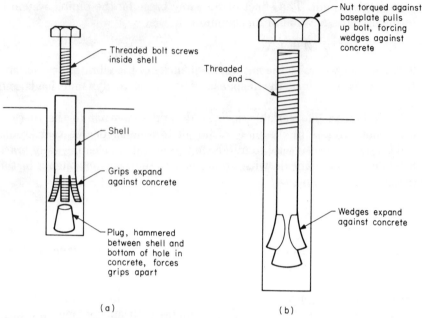

Figure 6.27 Anchor bolt types: (*a*) shell type; (*b*) wedge type.

(such as during retrofit), while embedded plates are used when they can be specified prior to pouring the concrete.

The strength of both types of plates is based on the shear cone capacity developed in the concrete by the applied tensile load, as shown in Fig. 6.26.

The strength of the bolts is a function of the embedment depth, the bolt or stud head diameter, the concrete strength, and the spacing between adjacent bolts (because of strength reduction due to overlapping of adjacent shear cones). Anchor bolts are installed by drilling a hole through the concrete into which the bolts are inserted. Depending on the type of bolt (shell or wedge, see Fig. 6.27), the bolt expands to grip the concrete either by hammering the bolt or by torquing the nut against the baseplate.

Embedded plates are anchored by headed studs which are welded to the plate and set in place prior to pouring the concrete. The added diameter of the head, as well as the greater adhesion of the concrete to the steel, gives the studs significantly higher capacities over the anchor bolts. Therefore embedded plates are preferred in all cases, except when the concrete has already been poured and embedded plates cannot be installed. In that case the only feasible alternative would be to provide a baseplate using concrete anchor bolts.

Baseplate size is based on the applied loading, required bolt spacing to develop shear cones fully, minimum required edge distance from bolt holes, prying effects due to plate flexibility, and physical space constraints. The matter of sizing baseplates is discussed more fully in Chap. 8.

6.7 Multiple-Pipe Restraint Frames

Most of the examples discussed so far in this chapter have concerned single-pipe support assemblies. When several pipes to be supported are located together, but not near the building steel, special pipe support structures must be built. Steel frames such as those shown in Fig. 6.28 are used to "gang-restrain" several pipes with a common structure. Multiple-pipe support frames may be used when pipes are running parallel to each other in close proximity. The multiple-pipe frames are an economical means of supporting several pipes by allowing for common major support steel and similar attachment methods. Frames may be attached to either

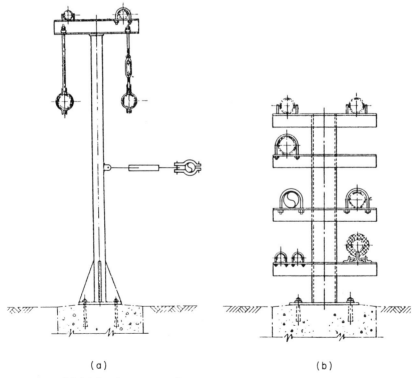

Figure 6.28 Multiple-pipe support frames.

the floor or the ceiling. It is recommended that pipes being supported on a common frame share the same code class or design criteria.

Frames are usually constructed by assigning the major support steel to be supplied with one of the pipes during construction. By supplying the majority of the pipe support steel with the first pipe, the remaining pipes can be easily attached to this existing steel.

When a multipipe frame is designed, it is recommended that loads be balanced on the structure. The engineer must identify and design for the most critical combination of all the individual pipe loads. Different load combinations may be critical for different sections to the frame.

Chapter 7

Piping Support Design Process

7.1 Introduction

The design, engineering, and construction of a power or process plant involve a long series of steps leading from conceptual study to turnover to client, as depicted in the sample power plant engineering and construction schedule shown in Fig. 7.1. In turn, each of these individual tasks comprising the total job must also be subdivided into a series of steps. Piping system support design is no exception; it, too, is subdivided into manageable portions. This task must be performed in parallel with the work of other disciplines, since each segment of the design process affects all others.

To ensure that a schedule is successfully met, the design process must begin far in advance of construction dates. The piping system supports may originally be designed conceptually in order to allocate space envelopes and estimate required quantities and sizes of steel and engineered hardware. Far prior to the detail design phase, the engineers must estimate restraint reaction loads acting on the building structure. Next, design criteria must be determined, and design standards for support arrangements, structural steel design, hardware selection, etc., should be developed to reduce engineering work and to identify any potential areas of concern as early as possible.

The design steps for nuclear, fossil fuel, petrochemical, shipboard, and industrial piping are all quite similar, even though they may vary as far as schedule and level of detail are concerned. Therefore, it is believed that this text may serve as a general guide for designers in almost any piping-related industry, after they have reviewed its contents and adapted it to the specific needs of the industry.

This chapter outlines the essential steps, whether sequential or parallel,

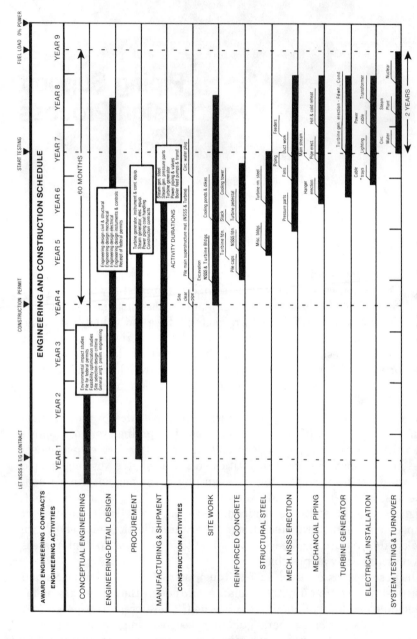

Figure 7.1 Power plant engineering and construction schedule. *(Courtesy of Engineering Planning and Management, Inc.)*

TABLE 7.1 Pipe Support Design Checklist

1. Have proper codes and design criteria been used? ☐
2. Have the design loads been determined and specified? ☐
3. Has frictional load been included in restraint design? ☐
4. Have thermal movements and rotations of pipe been considered in design? ☐
5. Have formulas been used correctly and referenced properly? ☐
6. Have all required references been shown properly? ☐
7. Have calculation number and mark number been indicated on all sheets properly? ☐
8. Has each page been signed and dated by designer and checker? ☐
9. Has all engineered hardware been correctly selected for the load and application? ☐
10. Has maximum swing of rod, strut, or snubber been limited to $\pm 6°$? ☐
11. In trapeze design, has each rod or strut been designed for 100% of total load? ☐
12. Have sketches been prepared for all nonstandard items? ☐
13. Are all steel stresses within allowable limits? ☐
14. Has proper edge distance been maintained for holes in baseplates and structural steel? ☐
15. Do all welds comply with ASME, Subsection, NF, App. XVII and meet minimum weld requirements (or AISC where applicable)? ☐
16. Have all welds been properly designed and indicated on sketch (use standard forms provided for weld design)? ☐
17. Has it been verified that there is no welding across the flanges of existing steel members? ☐
18. Has embedded plate been properly designed according to applicable procedure? ☐
19. Have anchor bolts been designed properly for maximum loads, and has correct factor of safety been used? ☐
20. Has the distance between anchor bolts and other embedded items been maintained as required? ☐
21. If not, has the bolt been derated as per catalog reference? ☐

that are necessary for piping system support design. The piping engineer may find it helpful to maintain a checklist of the various calculation steps used in designing supports. A sample checklist is given in Table 7.1. A detailed description of the design steps is summarized in the following section with the required task presented in the sequence of steps normally encountered in design. The support designer should bear in mind that there is no standard method of design. Each procedure must be individually tailored to specific project and industry needs. What is presented here is a sample intended to serve merely as a guideline for developing individual procedures.

7.2 Preparation of Information and Data Required for Pipe Support Design

The pipe support design function should not be initiated until the necessary reference data is compiled. This should include, as a minimum, the following:

1. Design documents, such as applicable piping codes, departmental

design standards, project design criteria, load capacity data sheets, and catalog material from the vendor supplying the hardware, company and industry design standards, etc.
2. Piping drawings showing the piping routing in plan and elevation. The drawings required are not only those of the piping to be restrained, but also those showing other piping in the area which may cause interference.
3. Isometric drawings indicating pipe routing in three dimensions.
4. Structural drawings indicating existing building structural steel, embedded steel, or concrete where anchor bolts may be attached. Sample steel drawings are shown in Figs. 7.2 and 7.3, respectively. Figure

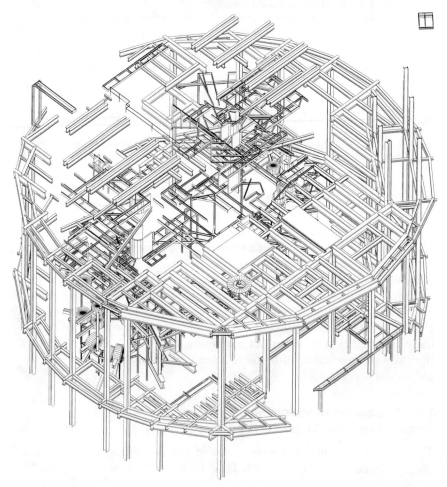

Figure 7.2 (a) Typical overhead steel drawing; (b) typical platform support structural

6.1 is another such drawing, showing typical support steel beams in plant arrangement.

5. Drawings indicating locations of equipment designed by other disciplines, such as electrical raceways; heating, ventilation, and air conditioning (HVAC) ductwork; and instrumentation and control lines. Usually, general arrangement drawings are useful in determining the locations of major equipment, vessels, etc.
6. Line list, which includes pipe schedule, service, pressure, temperature, code classification, insulation requirements, piping material, and fluid contents.

7.3 Selection of Pipe Support Locations

Once the required data has been assembled, the engineer may begin to spot preliminary pipe support locations. The first step is to establish a desired span for supporting each pipe size. This may be based on the standard weight span or on a lateral span determined by frequency or stress criteria (see Chap. 5 for determination of support spans) when seismic or wind loads are applicable. It is usually advantageous to develop a single span to satisfy both criteria. For example, if the weight span for a 4-in

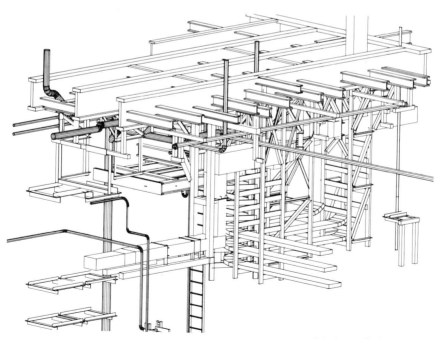

steel detail. *(Courtesy of Construction Systems Associates, Inc., Marietta, Ga.)*

(100-mm) line is not to exceed 14 ft (4.3 m) and the desired lateral span is determined to be 16 ft (4.9 m), a common weight-lateral support span should be chosen at the minimum, or 14 ft (4.3 m). If the lateral span for a 3-in (80-mm) line [weight span = 12 ft (3.7 m)] is found to be 25 ft (7.6 m), the support span should be chosen as 12 ft (3.7 m), with a lateral support located every 24 ft (7.3 m) or at every other weight support. It is usually more economical to combine functions at a single location (and therefore reduce total material cost and construction time) than to eliminate a few supports by "stretching out" the span between them and thereby staggering them with the other functions. The support span may also be reduced, so lines of many sizes running in parallel proximity may share common supports.

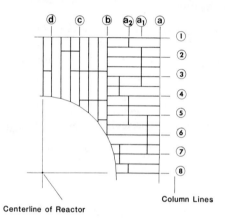

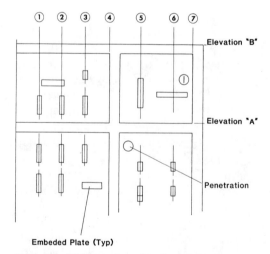

Figure 7.3 (*a*) Typical overhead steel arrangement; (*b*) typical wall steel drawing.

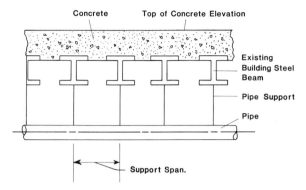

Figure 7.4 Recommended economical support location.

Once the desired span for the pipeline has been determined, the engineer should overlay or superimpose the piping drawings onto the plant structural steel drawings. At this point, preliminary support locations may be chosen. The locations selected should be at or close to existing steel and other points convenient for the attachment to the existing structure, and the locations should limit pipe lengths between supports to the predetermined span. The supports should preferably fall on the following locations:

1. On straight pipe, rather than on piping components, fittings, or bends
2. At locations which will not interfere with required maintenance
3. As close as practical to heavy load concentrations such as vertical runs, valves, branch lines, flanges, etc.
4. Near changes in the direction of the pipe run

It is nearly always preferable to reduce the span between supports where necessary to simplify the design as opposed to taking full advantage of the allowable span and being forced to bridge between existing building steel. This concept is illustrated in Figs. 7.4 and 7.5, which show economical and uneconomical locations, respectively. Conversely, sufficient distance to the structure to permit construction of the support must be maintained where necessary. As seen in Fig. 7.6, a wall penetration may be an ideal location for a lateral piping support (since the loads could be transmitted by lugs and no moments need be transmitted to the structure), but it is an impractical location for a lateral snubber, owing to space constraints.

When pipe support locations are chosen, the engineer must make full use of the drawings generated by other disciplines in order to avoid physical interferences with HVAC, electrical, instrumentation and control, mechanical, or structural space requirements. Composite drawings (see

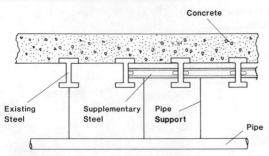

Figure 7.5 Not recommended support location.

Fig. 7.7) or plant scale models (see Fig. 7.8) are commonly used to prevent such interference. It is recommended that multidiscipline common supports be avoided. The various disciplines' equipment may not be installed at the same time, probably will not be installed by the same labor crafts, and will almost certainly be designed to different codes and criteria.

Preliminary support locations should be chosen with care, for once the stress analysis of the pipe has been completed, the supports often may not be moved without causing significant changes in pipe stress levels and support loads, resulting in a need for reanalysis. Therefore it cannot be overemphasized that the preliminary location of supports is one of the most important steps in the piping support design process.

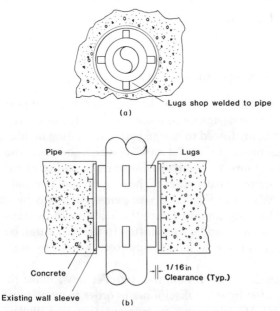

Figure 7.6 Two-way pipe support located in wall penetration: (a) elevation; (b) plan.

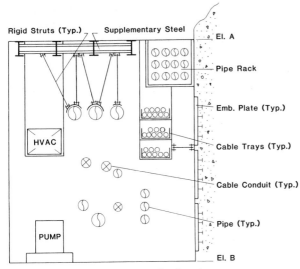

Figure 7.7 Typical area composite drawing.

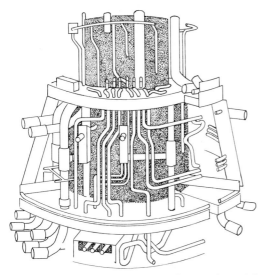

Figure 7.8 Typical nuclear power plant, scale model.

7.4 Marking Support on Isometrics

The next step in the design process is to mark the location and prospective direction of action for each support on the piping isometric drawings. The direction of action or function required can usually be determined in advance of the stress analysis calculation based on the design criteria for the system. For example, a system experiencing only normal operating loads would most likely require only vertical downward (gravity) supports, while a system designed to withstand wind loading, earthquake, or fluid hammer loads would require supports in the horizontal plane.

It must be reemphasized that prior to the pipe stress analysis, both the locations and, to some extent, the direction of action of the supports have a large degree of latitude. However, once the stress calculations have been completed, it is difficult to make changes in the support scheme. Deviation tolerances are usually established by the stress analysis engineer. For instance, location tolerances such as twice the pipe diameter or 12 in (30 cm) may be established as the maximum. Therefore the engineer should be relatively certain that the supports called for on the isometric drawing (as shown in Fig. 7.9) have sufficient space envelope and a convenient method of attachment to the building structure.

The completed isometric drawing is transmitted to the stress analyst to serve as the basis for the stress analysis calculations.

7.5 Pipe Stress Analysis

The pipe stress analyst uses the marked-up isometric drawing prepared by the support designer as input for the pipe stress calculation. The stress analyst must determine support configurations such that the piping will not be overstressed for any specific load combinations, as defined by the applicable code equations (see Chap. 4 for piping code equations). The marked-up piping isometric drawing is used to indicate the locations and types of supports which may be considered by the pipe stress analyst.

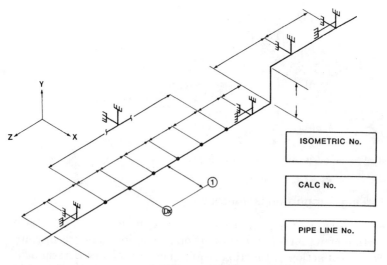

Figure 7.9 Typical isometric with initial pipe support locations.

The stress analyst must balance the need for pipe support with the need for sufficient flexibility to accommodate thermal expansion. The ideal solution would be to furnish the absolute minimum number of supports actually needed, with locations and functions chosen with the intention of preventing thermal lockup. Therefore, a large number of support locations and directions of action indicated on the isometric may go unused. The "optimum" use of pipe supports is realized by taking advantage of allowable piping stresses (per the applicable code) to minimize pipe support locations.

The stress analyst, besides determining the required locations and direction of action of the supports, also chooses the function. This leads to the second area (besides minimizing the number of supports) in which the analyst can provide an economical design.

The function of the support (i.e., rigid support versus constant spring, variable spring, or snubber) is normally dictated by the type of loading and thermal movement in the support direction. When there are relatively small vertical thermal movements, usually weight loads may be supported by rod hangers; when large thermal movements must be accommodated, a spring might be required. Likewise, a lateral support may be rigid when thermal movements are low, but a snubber (for shock loadings) may be required when the movements are large. Since rigid support components are less expensive than springs or snubbers, the stress analyst can save significant expense by minimizing and directing thermal growth. The use of snubbers may be a quick solution to the stress analysis problem; however, because of the high purchase price and maintenance cost, snubbers are a very costly solution and should be minimized.

Thermal movement may be directed by judiciously choosing the locations of the first rigid support on a run prior to performing the thermal expansion analysis (for guidance on finding the location of the first rigid support, refer to Chap. 5). By imposing zero movement at set locations, the analyst may permit the use of other rigid supports between these points.

Once the rigid support locations have been chosen, the analyst performs a "free thermal" calculation aimed at determining expansion stresses and movements in the pipe. This calculation considers the effect of only the rigid supports, leaving the pipe "free" of the effects of nonrigid supports. In many cases, the free thermal analysis is done prior to location of the rigid supports; these rigid supports are then placed at points of small thermal movement. The movements determined during the pipe free thermal calculation will be used by the support designer in selecting springs and snubbers.

The next analyses done by the pipe stress engineer are the restrained loading cases, where the supports (rigids, springs, and snubbers) are all considered rigid for their applicable loading case. The pipe loads thus calculated become the weight and occasional loads for the rigid supports, the weight loads for the springs, and the shock loads for the snubbers.

The pipe stress analyses (previously mentioned) may be performed either by computer or manual calculations. Manual calculations normally offer less accurate results than computer calculations and usually become less acceptable as the loadings become more complex. The advantage of manual calculations is that they may be less expensive and, when accuracy is not required, less time-consuming than computer calculations.

7.6 Determination of Support Design Loads

Once the piping system has been analyzed for weight, thermal, and occasional loads, the support loads determined by separate analysis must be combined to determine the resultant design load for each support. The required combinations are usually specified in the project design criteria. However, a rule of thumb is that all loads which can occur simultaneously should be considered simultaneously to determine the maximum and minimum forces acting on the support.

For example, since an earthquake is postulated to strike a piping system equally in the plus or minus direction and may occur when the system is operating or cold, the following loading combinations should be considered [DW = deadweight, TH = thermal load, EQ = earthquake (seismic)]:

DW + TH + EQ
DW + TH − EQ
DW + EQ
DW − EQ

From these four loading cases only the maximum loading in each direction would be selected to design the support.

The design loads are usually summarized on a unique data sheet for each support. A typical load and movement summary sheet is shown in Table 7.2. Referring to this table, note that the loads and movements may be presented in axes corresponding to either the local pipe axes or in-plant global axes (x, y, and z). The local axes correspond to the axial, vertical, and lateral directions of the pipe. Included on the summary sheet are also the pipe stresses calculated for each loading case for use in designing the piping attachments. Note that the same stress was calculated for deadweight plus seismic and deadweight minus seismic; this is because the piping codes require absolute summation of the stresses present during the individual loading cases.

When conditions result in more than one value for a particular design

TABLE 7.2 Sample Support Load and Movement Summary Sheet

Mark no: MS-R-43 Stress Iso: 4733
Function: Lateral support Node pt.: 27
System: Main steam Stress calculation: 341 Rev 1

Load case	Force			Deflection			Stress, psi (kPa)
	$F(x)$	$F(y)$	$F(z)$	Δx	Δy	Δz	
Thermal (TH)— Start-up	—	—	1,543	−1.03	0.57	0.0	1,122 (7,736)
Thermal— Operation	—	—	4,251	−2.97	1.51	0.0	3,113 (21,464)
Deadweight (DW)	—	—	−1,250	0.02	−0.07	0.0	1,237 (8,529)
Seismic	—	—	±12,110	±0.08	±0.29	0.0	±8,429 (58,118)
DW + seismic	—	—	10,860	0.10	0.22	0.0	9,666 (66,647)
DW − seismic	—	—	−13,360	−0.06	−0.36	0.0	9,666 (66,647)
DW + TH (max.+) + seismic	—	—	15,111	−2.87	1.73	0.0	—
DW + TH (max.−) − seismic	—	—	−11,817	−3.03	1.15	0.0	—

loading case, the maximum value in each direction is usually chosen for design purposes. This frequently occurs with thermal loading conditions which may represent several different operating conditions, such as start-up, normal operation, shutdown, and accident.

Once the support loads and movements are determined, the conceptual and detail design processes, as discussed in the following sections, will begin.

7.7 Pipe Support: Conceptual Design

A conceptual design sketch is prepared by the support designer while keeping in mind the number and type of support functions at a given location. The immediate concern is to select the type of hardware to be used based on a review of the plant area and the availability of supporting structures. The conceptual design may be prepared from design drawings, from a plant model, or through direct observation at the construction site. Although a conceptual design may have been considered during the location selection and isometric marking phase, now that the functions and the actual support loads are known, the intended designs should be reconfirmed. Additionally, much time may have passed since the conceptual design was prepared, so it must be reviewed in light of the plant changes which may have taken place in that interval.

During the conceptual phase of design, the pipe support arrangement and hardware to be used at the specified location are determined. Any other piping or equipment that could interfere with the resulting design should be identified with their locations relative to the support under study. A typical conceptual design is shown in Fig. 7.10.

Every effort should be made to use standard pipe support hardware for which load capacity data sheets are available from the vendor, thus eliminating the need for some of the analysis. Structural steel frames which are difficult to analyze should be avoided, except when absolutely necessary or when construction savings (i.e., combining many supports into a single frame) justify the additional analytical expense. Welded pipe attachments such as lugs, trunnions, etc., should be avoided in favor of clamps, in order to minimize construction time and pipe wall stresses induced by the load on the attachment. Clearances should be left in the frame sufficient to permit thermal movements, and support steel should be kept clear of walkways and operators' areas.

The general rule of thumb is to select the simplest design or hardware possible for each support. A simple design usually has a minimum of hardware and does not require welding to pipe. Adjustability in the hardware such as with clamp and strut arrangements should be given consideration.

When the piping system requires *in-service inspection,* the welds on the pipes will require periodic inspection; therefore support hardware must be located far enough away to allow access to these welds. Figure 7.11 illustrates the closest recommended location for placing a support

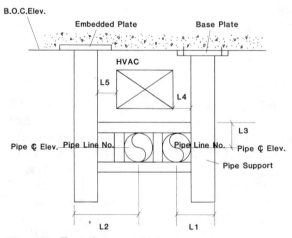

Figure 7.10 Typical conceptual design.

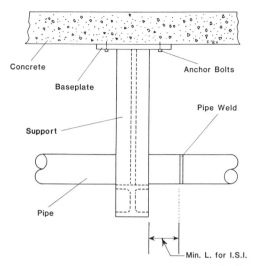

Figure 7.11 In-service inspection space required.

near a weld needing inspection. The in-service inspection minimum clearance distance is defined in Section XI of the ASME code and is usually dependent on the pipe wall thickness, such as

$$L_{min} = 3T + 2 \text{ in } (5.1 \text{ cm})$$

where T = pipe wall thickness (in or cm). This clearance dimension should be complied with in selecting locations of pipe supports.

7.8 Pipe Support: Detail Design

Once the conceptual pipe support arrangement or design has been selected, the support structural members are selected and sized. Steel selection may be either a simple or complex task, depending on several design considerations. For example, the steel structures must support the imposed load, fit within the available space limitations, simulate the spring constant assumed for the support in the stress analysis calculation, appeal to aesthetics, and be adequate from a pipe wall local stress perspective.

The design loads are imposed on the steel structure at one or more locations, either through support hardware or directly through the bearing of the pipe wall against the frame. These pipe loads are transmitted to the building structure through the pipe support. Thus the steel members and their joint connections must be able to safely withstand the applied loadings. Therefore a structural analysis of the support must be performed to substantiate that the supporting structure is adequate for the imposed loads. This is achieved by demonstrating that the actual stresses in the steel are below those permitted by the applicable code, whether ASME, Section III, Subsection NF (for nuclear piping) or the AISC specification (for nonnuclear piping). The analysis required may range from use of simple beam equations to computer analysis. This procedure is described in Chap. 8 in detail, with examples.

Structural design of support steel is normally the most time-consuming part of the design process. Time may be saved by eliminating this step through the use of vendors' standard load-rated hardware. Additional engineering time may be saved through use of standard or generic designs which have been preengineered and rated for load.

Pipe support spring constants are used to simulate support stiffness when a computerized stress analysis of the piping system is performed. Most programs, even when modeling a "rigid" support, model the support with a finite stiffness in terms of pounds per inch (newtons per meter) or foot-pounds per radian (newton-meters per radian). In the first-pass stress analysis, the stress analyst will use the value specified by the project design criteria. The values chosen for the support stiffnesses influence the load distribution during static loading cases and the system natural frequencies (and thus the loadings) during dynamic loading cases. To accurately simulate the response of the piping system under various loadings, the actual stiffnesses (spring constants) of the installed supports should be close to the assumed stiffnesses (at least relative to the stiffness of the pipe).

The design spring constant may be presented either as an absolute targeted spring constant or as a maximum support displacement under load. Typically, it is desirable to have a support sufficiently "rigid," for example, above 1×10^6 lb/in (1.8×10^8 N/m). When one is designing to a maximum displacement, the effect is to increase the spring constant as the load increases. Assuming that load increases with pipe diameter, this will ensure that the larger lines receive more rigid supports, while those on smaller lines will be more flexible. This avoids the need for massive supports for small lines, while still maintaining the "rigidity" of the support *relative* to the pipe.

A *support spring constant* is defined as the ratio of the applied force to deflection or the applied moment to rotation at the point of loading. This is represented by

$$K = \frac{F}{\Delta} \quad \text{or} \quad K = \frac{M}{\theta}$$

where K = spring constant, lb/in or ft·lb/rad (N/m or N·m/rad)

F = imposed force, lb (N)

Δ = resulting displacement, in (m)

M = imposed moment, ft·lb (N·m)

θ = resulting rotation, rad

Often the stiffness or the spring constant (one term being the inverse of other) is the governing criterion in the sizing of the steel, rather than the allowable stresses. Further discussions of calculation methods with illustrative examples are included in Chap. 8.

The pipe support steel should also be a minimum size from an aesthetic point of view. In engineering design applications, a design should look "correct" or pleasing as well as inspire confidence in the design. Suggested minimum structural steel member sizes are presented in Table 7.3. The table presents pipe size ranges and corresponding minimum steel sections, based on industry practices. These sizes are by no means guaranteed to withstand the applied loads; however, they may be considered as the starting point in the support design.

Pipe wall local stresses are developed by pipe loads acting at the attachment of the pipe to the steel. Stresses are induced both by welded attachments (lugs, trunnions, saddles) and by the line load caused by the pipe resting against, pressing against, or striking the steel. The local stress is normally dependent on the pipe radius, wall thickness, load magnitude and type, and attachment size. Normally the larger the attachment, the lower the induced stress will be. Therefore the support steel must be sized to keep the induced stresses below allowable levels.

The allowable induced stress is the allowable pipe stress, as per the applicable piping code, minus the actual stress in the pipe, as calculated by the pipe stress analyst. The remaining margin is available for use in designing the pipe attachments and line loads. These stresses are discussed further in Chap. 8.

TABLE 7.3 Aesthetic Steel Member Size

Pipe size (nominal diameter), in (mm)	Structural steel member minimum size	
	in	mm
0.5–2 (15–50)	L $2 \times 2 \times \frac{1}{4}$	L $50.8 \times 50.8 \times 6.35$
2.5–4 (65–100)	L $3 \times 3 \times \frac{3}{8}$	L $76.2 \times 76.2 \times 9.53$
5–8 (125–200)	W 4×13, TS $4 \times 4 \times \frac{1}{4}$	M $102 \times 102 \times 19$
10–14 (250–350)	W 8×17, TS $6 \times 6 \times \frac{3}{8}$	UB $203 \times 133 \times 36$
16–20 (400–500)	W 8×24, TS $8 \times 8 \times \frac{3}{8}$	UB $203 \times 165 \times 36$
22–30 (550–750)	W 10×33, TS $8 \times 8 \times \frac{3}{8}$	UB $254 \times 203 \times 49$
32–up (800–up)	W 12×40, TS $10 \times 10 \times \frac{1}{2}$	UB $305 \times 254 \times 79$

L = angle, W = wide-flanged beam, TS = tube section; M = wide-flanged beam, and UB = tube section (European).

Once the pipe support structure has been analyzed, the engineer will have determined reactions at the attachments to the main structure. The capacity of these attachments to the structure must be verified. If the attachment is to building steel, the connecting weld must be designed; if the attachment is to an embedded plate or a baseplate with concrete anchor bolts, the capacity of the plate and bolts must be checked. Examples of these calculations are given in Chap. 8.

The reactions transferred to the structure, once determined, must be transmitted to the structural engineer responsible for the structural integrity of the building. These support reactions, sometimes called *footprint loads,* may be transmitted on an individual or an area basis. At this point it would be the responsibility of the structural engineer to verify that the actual footprint loads fall within the original assumptions of pipe loadings used to design that section of the plant.

7.9 Detail Design Drawing and Bill of Materials

Once the pipe support design is completed, a detailed design drawing must be prepared. This drawing will become the basis for fabrication and installation as well as the permanent record of the design.

A sample drawing is shown in Fig. 7.12. As noted, the drawing should include sufficient information to uniquely identify the support, should reference the design inputs, and should give adequate details required for fabrication and installation. The drawing in Fig. 7.12, includes such identifying information as the support mark number, the line number, the stress analysis calculation number, and reference drawing numbers. The support location is identified through the location plan and the pipe centerline elevation. The design information is specified by the arrangement showing all materials as well as fabrication details such as welds. All materials are called out in a bill of materials which, besides clarifying the design information, facilitates material takeoffs to estimate construction costs. The detail drawing should incorporate construction tolerances. In a typical detail drawing, a gap in the direction of support is usually shown as $\frac{1}{16}$ in (1.6 mm); traditionally, a gap of up to $\frac{1}{16}$ in (1.6 mm) is considered full restraint, since flush fit of pipe and steel is difficult to achieve during construction.

7.10 Checking

Since the design and installation of pipe supports is a costly activity, and since failure of a pipe may be still more costly, it is recommended that the design inputs, support calculation, and detail drawing be independently

Piping Support Design Process

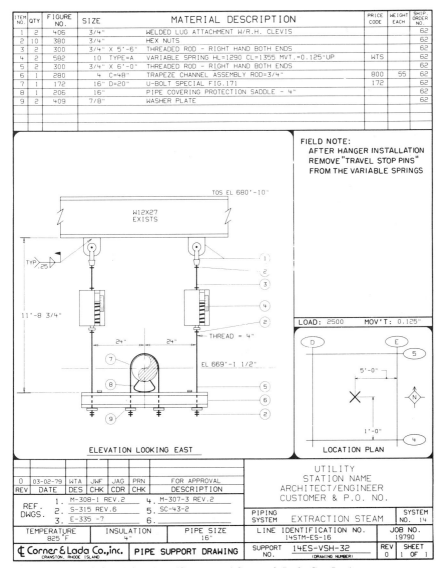

Figure 7.12 Final detail drawing. *(Courtesy of Corner & Lada Co., Inc.)*

checked prior to release for fabrication and installation. In the case of safety-class nuclear pipe supports, it is required by law that the work be checked by an independent verifier. The verification may be conducted either by direct checking of the originator's work or by the generation of similar results using different calculation methods.

7.11 Fabrication, Installation, and Verification of As-Built Configuration

Since drawings are rarely a perfectly accurate representation of the real world, some pipe supports are fabricated and installed with deviations from the design drawing.

When changes must be made because of existing field conditions, it is, of course, preferable to provide a better, or stronger, design. This would entail substituting stronger members, providing stronger welds, or using shorter member spans. Of course, this is not always possible, and many changes will have a detrimental or at best ambiguous effect on the support and piping system. For example, when a support location must be changed, a movement in one direction may reduce thermal load but increase weight loading, and vice versa. For this reason, it is recommended that an experienced support engineer and stress analyst be available for consultation during construction.

Subsequent to installation, any deviations should be noted on the drawings, and the revision block should be marked *as-built*. An engineer should evaluate whether any of the deviations are major; if so, the calculations should be reviewed to determine the capacity of the as-built configuration. In cases where the locations, directions, functions, or spring constants of any restraints change significantly, it may be necessary to redo the stress analysis calculation, thus generating new design loads for all restraints in the system.

Chapter

8

Manual Calculation Methods

8.1 Introduction

The previous chapters have emphasized the determination of pipe support design loads, the conceptual design of the pipe support, and the selection of proper hardware. Once these steps have been completed, the next stage of the design process is the sizing of the structural components, such as steel members, baseplates, and pipe attachments. Manual calculation methods used to analyze these items are described in this chapter.

Note: The reader should be warned of the danger of considering the material and sample calculations presented here as an interpretation of the code requirements. A careful study of the applicable code, which may be procured from the ASME or AISC, is recommended. It must be emphasized that a competent engineer must state the applicability of all design examples presented in this chapter.

Manual calculations are usually an approximation of actual structural response, since simplifying assumptions are usually made to save effort. In most cases, manual calculations yield reasonably accurate results for the evaluation of simple structures. However, in cases involving highly complex structures, dynamically applied loads, flexible baseplates subjected to prying forces, or unorthodox pipe attachments, a computer-assisted analysis should provide greater accuracy and may even save engineering and material costs. Computerized calculation methods are discussed in Chap. 9.

The analysis required to determine the structural capacity of a pipe support can usually be divided into four distinct areas: calculation of local stresses developed in the pipe wall due to attachments, determination of capacities of vendor-supplied hardware, structural analysis of support steel and connections, and evaluation of the strength of attachments to the main building structure.

Methods of analysis are presented in the previously mentioned order, partially for convenience and partially because the engineer is advised to pursue this order when analyzing a support. To ensure that all parts of the support are checked, analysis should proceed from one end of the support (the pipe attachment), through the support configuration (hardware, steel, and welds), to the building attachment joints (baseplates, welds, etc.). Only in the simplest support arrangement, or with a very experienced designer, may the weakest links in the support chain be determined by inspection. Therefore the engineer should proceed through the analysis in an orderly, thorough manner, ensuring that all components of the support are verified as capable of withstanding their applied loads.

The rest of this chapter is divided into three sections. Section 8.2 presents background information about strength of materials and statics. Because of space limitations, this book cannot include a full discussion of structural design or strength of materials. For this, the reader is directed to seek out a text devoted to this subject. The authors assume that the reader has a working knowledge of these subjects; therefore, only the basics are presented.

Section 8.3 deals with the design equations necessary to perform the four analysis steps listed. Section 8.4 presents examples demonstrating use of the methods outlined here.

8.2 Strength of Materials

Before we introduce the design equations, we offer a refresher on the principles of strength of materials.

8.2.1 Stress-strain relationship

Stress is the internally distributed force (given in terms of force per unit area) which permits a material to resist externally applied loads. Stress may be of two types: *normal* (either tensile or compressive), which resists the longitudinal elongation or compression of the body, and *shear,* which resists the relative slippage of adjacent cross sections of the body. For further discussion of material failure modes under various stress types, see Chap. 4.

Stress is always accompanied by material deformation. These deformations may be translational or rotational. The ratio of axial deformation to total member length (Δ/L) is known as *strain* ϵ. Hooke's law states that, for most materials, strain is linearly proportional to the stress induced, up to the level known as the *yield point.*

The yield point of a material is determined by applying a uniaxial tensile load to a bar of constant cross section and plotting the strain versus stress as the load is increased. The resulting curve will show an area of linear proportionality and an area of unrestrained deformation (see Fig. 8.1). The point dividing these two areas is known as the *yield stress*. However, since this "point" is rarely clearly identifiable, the yield point is normally taken to be that stress which leaves a 0.2 percent residual strain once the loading is removed.

The slope of the linear portion of the curve is known as the *modulus of elasticity* E, or the unit stress per unit strain. Measured in the tension test, the modulus of elasticity is

$$E = \frac{PL}{A\Delta} = \frac{\sigma}{\epsilon} \quad \text{psi, (N/mm}^2\text{)} \tag{8.1}$$

where P = applied load, lb (N)
  L = length of specimen, in (mm)
  A = cross-sectional area of specimen, in² (mm²)
  Δ = observed elongation of specimen under load, in (mm)
  ϵ = strain, in/in (mm/mm) (dimensionless)
  σ = normal stress, psi (N/mm²)
  E = modulus of elasticity, psi (N/mm²)

Both the yield stress and the modulus of elasticity vary with the temperature of a material. These values have been determined for many materials through experimental testing and are published by the American Society for Testing of Materials (ASTM).

8.2.2 Axial stresses

Axial stresses may be either tensile or compressive. Tensile stresses similar to those developed in the uniaxial tensile test are frequently developed in structural problems. These stresses tend to elongate the member

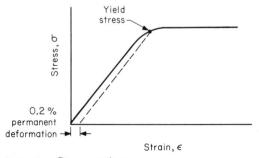

Figure 8.1 Stress-strain curve.

by pulling along an axis parallel to the length of the member. The most common application of tensile loads in the design of pipe supports is the hanger rod, which is, by design, loaded only with tension.

Tensile stresses are equal to the applied force divided by the area, or

$$\sigma_T = \frac{P}{A} \tag{8.2}$$

where σ_T = tensile stress, psi (N/mm^2)
P = tension force, lb (N)
A = cross-sectional area, in^2 (mm^2)

The failure mode for tensile stresses is usually material yield. Therefore the allowable tension stress is normally limited to the yield stress reduced by a safety factor.

Compressive stresses differ from tensile stresses in that the failure mode is instability rather than high stress levels. A compressive load on a column becomes eccentric and causes an overturning moment whenever the column is disturbed by lateral load. If the overturning moment exceeds the restoring strength of the column, the column will buckle. Therefore the engineer must be more careful when designing for compression loads than for tension. The compressive capacity of a column is dependent on its slenderness ratio, which is defined as

$$\text{Slenderness ratio} = \frac{Kl}{r} \quad \text{(dimensionless)} \tag{8.3}$$

where K = a constant dependent on boundary conditions
l = unbraced length of column, in (mm)
r = least radius of gyration of the member = $\sqrt{I/A}$, in (mm)
I = moment of inertia of cross section, in^4 (mm^4)
A = area of cross section, in^2 (mm^2)

Theoretical and recommended values of K for some typical column end conditions are shown in Fig. 8.2.

Typical values for the radius of gyration are published along with other cross-sectional properties for standard structural shapes. Formulas for cross-sectional properties of various types of members are given in Table 8.1. The values of r will vary with respect to the axis under consideration: therefore all axes of the member should be reviewed to select the appropriate value for the calculation. If the unbraced length of a member is the same in each direction, then the lowest value of r (for the weakest axis) will govern the design.

Columns are usually classified into two types. The first is the short compression member shown in Fig. 8.3a. A short compression member fails by crushing, although a ductile material will probably deform considerably before failure. The long column (Fig. 8.3b) fails through bending of the ductile material under an eccentric compressive load.

The classification of the column is dependent on the column slenderness ratio as well as the material properties. For low-carbon steel, a short compression member usually has a slenderness ratio below about 40. For aluminum and magnesium, any compression member with a slenderness ratio below about 10 is considered a short compression member. The column classification is made by comparing the slenderness ratio to the material parameter C_c, where C_c is defined as

$$C_c = \sqrt{\frac{2\pi^2 E}{S_y}} \tag{8.4}$$

where E = modulus of elasticity, psi (N/mm²)
S_y = material yield strength, psi (N/mm²)

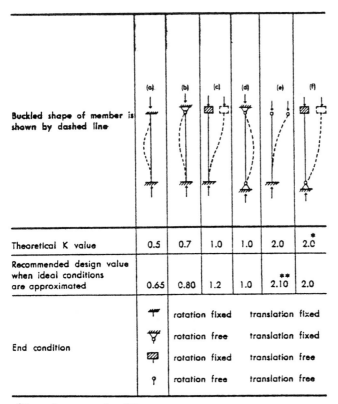

Figure 8.2 Effective length coefficient K for typical column conditions. SOURCE: Boiler and Pressure Vessel Code, Section III, Subsection NF (*courtesy of ASME*).

TABLE 8.1 Properties of Thin Sections Where Thickness (t) is Small, b = Mean Width, and d = Mean Depth of Section

Cross-sectional property	Section					
	⊤ (T)	H	□	⊏ (C)	⌐ (L)	○
I_x	$\dfrac{td^3(4b+d)}{12(b+d)}$	$\dfrac{td^2}{12}(6b+d)$	$\dfrac{td^2}{6}(3b+d)$	$\dfrac{td^3(2b+d)}{3(b+2d)}$	$\dfrac{td^3(4b+d)}{12(b+d)}$	$t\pi r^3$
S_x	$\dfrac{td^2(4b+d)}{6(2b+d)}$ bottom $\dfrac{td}{6}(4b+d)$ top*	$\dfrac{td}{6}(6b+d)$ *	$\dfrac{td}{3}(3b+d)$	$\dfrac{td}{3}(2b+d)$ top $\dfrac{td^2(2b+d)}{3(b+d)}$ bottom*	$\dfrac{td}{6}(4b+d)$ top $\dfrac{td^2(4b+d)}{6(2b+d)}$ bottom*	$t\pi r^2$
I_y	$\dfrac{tb^3}{12}$	$\dfrac{tb^3}{6}$	$\dfrac{tb^2}{6}(b+3d)$	$\dfrac{tb^2}{12}(b+6d)$	$\dfrac{tb^3(b+4d)}{12(b+d)}$	—
S_y	$\dfrac{tb^2}{6}$	$\dfrac{tb^2}{3}$	$\dfrac{tb}{3}(b+3d)$*	$\dfrac{tb}{6}(b+6d)$*	$\dfrac{tb^2(b+4d)}{6(b+2d)}$ right side $\dfrac{tb}{6}(b+4d)$ left side*	—

I_{xy}	0	0	0	0	$\dfrac{tb^2d^2}{4(b+d)}$	0
R	$\dfrac{t^2}{3}(b+d)$	$\dfrac{t^3}{3}(2b+d)$	$\dfrac{2tb^2d^2}{b+d}$	$\dfrac{t^3}{3}(b+2d)$	$\dfrac{t^3}{3}(b+d)$	$2t\pi r^3$
r_x max. or min.	$\sqrt{\dfrac{d^3(4b+d)}{12}}\Big/\sqrt{b+d}$	$\sqrt{\dfrac{d^2(6b+d)}{12(2b+d)}}$	$\sqrt{\dfrac{d^2(3b+d)}{12(b+d)}}$	$\sqrt{\dfrac{\tfrac{d^3}{3}(2b+d)}{b+2d}}$		0.7071r
Neutral axis	$\dfrac{d^2}{2(b+d)}$ down from top			$\dfrac{d^2}{b+2d}$ down from top	$\dfrac{d^2}{2(b+d)}$ down from top $\dfrac{b^2}{2(b+d)}$	
r_y min. or max.	$\sqrt{\dfrac{b^3}{12(b+d)}}$	$\sqrt{\dfrac{b^3}{6(2b+d)}}$	$\sqrt{\dfrac{b^2(b+3d)}{12(b+d)}}$	$\sqrt{\dfrac{b^2(b+d)}{12(b+2d)}}$		

* = Add t/2 to c for S.

SOURCE: *Design of Welded Structures*, by Omer W. Blodgett. (Courtesy of the James F. Lincoln Arc Welding Foundation.)

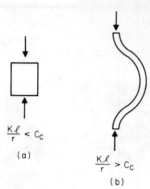

Figure 8.3 Compression failure modes: (*a*) short column (crushing failure); (*b*) long column (ductile failure).

The column is a short column if $Kl/r < C_c$. The allowable compressive stress (from the ASME and AISC codes) is given by

$$F_a = \frac{[1 - (Kl/r)^2/(2C_c^2)]S_y}{\frac{5}{3} + (3Kl/r)/(8C_c) - (Kl/r)^3/(8C_c^3)} \qquad \text{psi (N/mm}^2\text{)} \qquad (8.5)$$

The column is long when Kl/r is greater than C_c. In this design condition, the allowable compressive stress is given by

$$F_a = \frac{12\pi^2 E}{23(Kl/r)^2} \qquad \text{psi (N/mm}^2\text{)} \qquad (8.6)$$

8.2.3 Bending stresses

Normal stresses, besides being caused by axial forces, can also be caused by bending moments acting on the member cross section. *Bending moments* are internal member moments which resist externally applied moments in order to maintain the member in equilibrium. The magnitude of the bending moment at any point on a member is found by cutting an imaginary section at that point and solving the free-body diagram for equilibrium. Figure 8.4 shows the bending moment in a cantilever loaded at its end.

A bending moment diagram is a visual illustration of the bending moment value along the beam length. Bending moment diagrams are useful to the engineer because they point out locations of maximum bending moment. The diagram is generated by starting at one end of the beam and determining the moment at that location. One continues to find the bending moments at successive locations along the beam length by cutting sections and solving the resulting segment for equilibrium (summing the forces and moments). The moments calculated along the beam length are then plotted on the diagram. The bending moment diagram in Fig. 8.4 shows the maximum moment in the member to be located at the fixed end, with a magnitude of PL.

The bending moment is resisted by the cross section through normal stresses (tensile on one side of the neutral axis of the section, compressive on the other side). Stress due to bending may be calculated by the flexure formula. Bending stresses are usually far more significant than normal stresses due to axial forces; therefore the flexure formula in its many forms is one of the most commonly used equations in structural analysis.

The *flexure formula* states that the value of the bending stress σ_b at any point on the cross section of a member is:

$$\sigma_b = \frac{Mc}{I} \quad \text{psi (N/mm}^2\text{)} \tag{8.7}$$

where M = bending moment on the cross section, in·lb (mm·N)
c = distance from neutral axis to point of interest, in (mm)
I = moment of inertia of cross section, in^4 (mm^4)

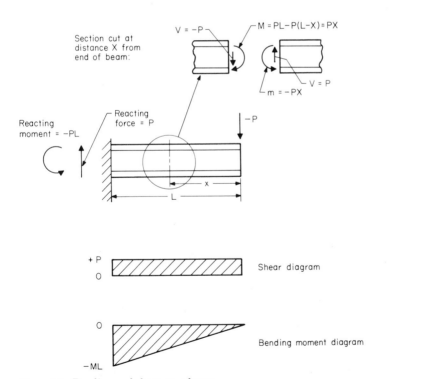

Figure 8.4 Bending and shear on a beam.

By inspection at the point where c is at its maximum, the bending stress is likewise maximized. This point is the *extreme fiber* of the section. At this location, the maximum bending stress is defined as

$$\sigma_{max} = \frac{M}{S} \quad \text{psi (N/mm}^2\text{)} \tag{8.8}$$

where

$$S = \text{section modulus} = \frac{I}{c_{max}} \quad \text{in}^3 \text{ (mm}^3\text{)}$$

A bending moment is usually not a destabilizing load, as is a compressive load, although under certain conditions portions of the cross section in the compression area of the bending moment may require checking for local buckling. Therefore the failure mode for bending is material yielding. For this reason the allowable stress for bending is usually limited to the material stress reduced by a safety factor.

Since both bending moments and axial forces create normal stresses, the effects of these two load types must be considered together. Most codes provide for an interaction formula to account for the combined effects.

8.2.4 Shear stress

Shear stresses are so named because the stress tends to shear, or cut, the member in two. These stresses resist the relative slippage of adjacent cross-sectional planes in the members and can be caused by either shear forces or torsional moments. Figure 8.4 shows a beam loaded with a shear force as well as a section cut showing the shear stresses on the cross section.

A shear diagram is a visual representation of the shear distribution along the length of the beam. Shear diagrams are useful to the engineer because they show the location of the maximum shear loads. The point where the shear force is equal to zero in a shear diagram often identifies a point of maximum or minimum bending moment. Thus shear analysis is conducted not only to determine maximum shear stress but also to locate the point of maximum bending moment, since bending moments can be predominant factors in member design.

Shear stress is rarely the governing design condition and therefore is usually neglected during the initial beam-sizing stage of design. However, once member sizes have been selected, the shear stress should be checked. Shear stress calculations are usually classified into those for open or closed section types due to the torsional capacities of the members. Open sections such as wide flanges, channels, and angles are not strong under torsional loading, while closed sections such as tubes and pipes are.

For open sections, the shear stress τ may be calculated as:

$$\tau = \frac{VQ}{It} + \frac{Tt}{R} \quad \text{psi (N/mm}^2) \tag{8.9}$$

where V = shear force on cross section, lb (N)
Q = static moment of area above point of interest = area above point of interest multiplied by distance between center of gravity of this area and center of gravity of entire section, in^3 (mm^3)
I = moment of inertia of cross section, in^4 (mm^4)
t = member thickness at point of interest, in (mm)
T = torsional moment on cross section, in·lb (mm·N)
R = torsional resistivity, in^4 (mm^4)

For a rectangular section, the $VQ/(It)$ term gives a parabolic stress distribution varying from zero at the edges to $1.5V/A$ at the center. This distribution has been considered in the safety factor for the AISC allowable shear stress on wide-flange webs, which need only be calculated as V/A.

The torsional resistivity may be approximated, for a rectangle, as

$$R = \tfrac{1}{3}bt^3 \tag{8.10}$$

where b = rectangle length, in (mm), and t = the rectangle thickness, in (mm). For an open section, the torsional resistivity may be approximated by taking the sum of the resistivities of the rectangles making up the section. Additionally, the AISC code provides the resistivities for wide flanges along with the other cross-sectional properties.

Closed structural sections have much more resistance to torsional loads than do open sections. Closed sections are usually classified as rectangular or circular, with circular being less commonly used for pipe supports (except in applications such as trunnions). For closed rectangular sections the expression for shear stress is approximated by

$$\tau = \frac{VQ}{2It} + \frac{T}{2bdt} \tag{8.11}$$

where t = wall thickness, in (mm)
b = mean width of section, in (mm)
d = mean depth of section, in (mm)

The shear stress on a circular cross section (such as a pipe) can be calculated by

$$\tau = \frac{VQ}{2It} + \frac{TD_o}{2J} \tag{8.12}$$

where D_o = outer diameter of section, in (mm)
J = polar moment of inertia = $\pi(D_o^4 - D_i^4)/32$, in^4 (mm^4)
D_i = inner diameter of section, in (mm)

Shear stress allowances have been historically determined experimentally and are normally given in codes as a function of the material yield strength.

8.2.5 Deflection

Since excessive deflection may cause design problems such as pipe sagging, unequal load distribution, etc., it is usually desirable to limit deflection in designs of the supporting structures. Calculation of deflections is also used to find support stiffnesses to be used as pipe stress input data. Therefore, a knowledge of how to determine deflection of various structures under loading is necessary. The pipe support deflection is usually determined at the point of load because this is the location for which the stiffness needed for the pipe stress analysis is found.

Deflections can be found once the relationship among shear, moment, rotations, and deflection is understood. The fact that points of zero shear indicate points of maximum and minimum bending moment demonstrates the derivative relationship of shear to moment.

The bending moment in the beam varies directly with the shear value times the length of the beam, as shown by

$$M_x = \int_0^x V \, dl + M_0 \tag{8.13}$$

where M_x = bending moment at point x along length of beam, in·lb (mm·N)
V = shear on cross section of beam, lb (N)
l = length of beam, in (mm)
M_0 = initial moment at boundary $x = 0$, in·lb (mm·N)

The bending moment divided by the product of the modulus of elasticity E and the section moment of inertia I is, in turn, the first derivative with respect to length of the angle of rotation of the loaded beam. The rotation is, in turn, the first derivative of the beam deflection:

$$\theta_x = \int_0^x \frac{M}{EI} \, dl + \theta_0 \tag{8.14}$$

$$\Delta_x = \int_0^x \theta \, dl + \Delta_0 \tag{8.15}$$

where θ_x = rotation at point x, rad
θ_0 = initial rotation, rad
Δ_x = displacement at point x, in (mm)
Δ_0 = initial displacement, in (mm)

Problem 8.1 These relationships are more easily understood when they are applied to an example. Figure 8.5a shows a simply supported beam of length l loaded in the center with a force P. From statics, it can be determined that the end reaction forces are $P/2$ and the end moments M_0 are 0. The shear diagram is shown in Fig. 8.5b.

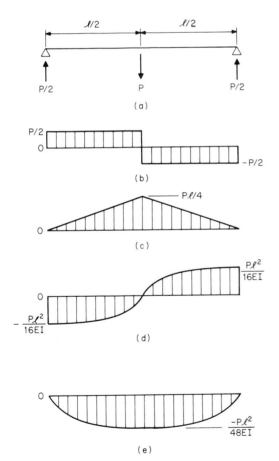

Figure 8.5 (a) Loading diagram; (b) shear diagram; (c) moment diagram; (d) rotation diagram; (e) displacement diagram.

The moment at any location is equal to

$$\int_0^x V\,dl + M_0$$

Since $M_0 = 0$,

$$M_x = \begin{cases} \dfrac{Px}{2} & \text{for } x \leq \dfrac{l}{2} \\ \dfrac{Px}{2} - P\left(x - \dfrac{l}{2}\right) = \dfrac{P(l-x)}{2} & \text{for } x \geq \dfrac{l}{2} \end{cases}$$

The maximum moment occurs at $x = l/2$, where $M = Pl/4$. The minimum moment occurs at $x = 0$ or $x = l$, where $M = 0$. An alternative method of solution to this problem is to plot the cumulative area under the shear diagram. This will yield the moment diagram shown in Fig. 8.5c.

Rotation may be found similarly at any location. However, the initial rotation at each support point must be considered. By symmetry, the rotation at each end is equal, although opposite in direction. At the center the rotation must be zero. Therefore,

$$\theta_{l/2} = 0 = \int_0^{l/2}\!\!\int \frac{V\,dl}{EI} + \int_0^{l/2} \frac{M_0\,dl}{EI} + \theta_0$$

For $M_0 = 0$ and $V = P/2$,

$$\theta_0 = \frac{-Pl^2}{16EI}$$

By symmetry,

$$\theta_l = \frac{Pl^2}{16EI}$$

By using the above formula and knowing the end rotations, the rotation may be determined at any location along the beam length. Alternatively, the rotation diagram (Fig. 8.5d) may be plotted by starting with the end rotation θ_0 and plotting the cumulative area under the moment diagram, divided by EI, from that point.

Displacement may be found by integrating the rotation. Since the supports are assumed to be "unyielding," the initial displacement $\Delta_0 = 0$. The displacement at the beam center point may be found by

$$\Delta_{l/2} = \int_0^{l/2}\!\!\int \frac{V\,dl}{EI} + \int_0^{l/2}\!\!\int \frac{M_0\,dl}{EI} + \int_0^{l/2} \theta_0\,dl + \Delta_0$$

For $\Delta_0 = 0$, $\theta_0 = -Pl^2/(16EI)$, $M_0 = 0$, and $V = P/2$,

$$\Delta_{l/2} = \frac{-Pl^3}{48EI}$$

Again, displacements may be solved for directly or found by plotting the cumulative area under the rotation diagram, as shown in Fig. 8.5e.

For most typical beam configurations, the shear, moment, rotation, and deflection equations have already been developed for easier use. Some of these formulas are shown in Fig. 8.6.

8.3 Support Analysis

As mentioned previously, the support review process usually consists of four stages: the local stress evaluation, vendor hardware review, structural steel analysis, and evaluation of building attachments.

8.3.1 Local stress evaluation

The first stage of the support review process is usually the evaluation of local stresses in the pipe wall. Normally the forces and moments carried in the pipe are distributed uniformly through the cross section. When the pipe is supported, however, these loads are usually removed over a reduced area and through discontinuities. This induces additional stress in the pipe wall in the area of the pipe attachment. The total pipe wall stress—or the sum of the pressure stress, the moment stress, and the local stress—must fall within the stress limits specified by the applicable piping code (see Chap. 4).

Numerous methods are available for estimating the induced stresses in the wall due to pipe attachments. Since this is a shell-bending problem, the most accurate solution method is finite-element analysis by computer. This, however, becomes prohibitively costly when it is done for every welded attachment or line load. Therefore, approximate methods have been developed to estimate the locally induced stresses, some of which are presented here. The engineer may use other methods if the model can be justified for the actual case.

Pipe attachments are classified into two types: nonintegral, where the attachment is not welded to the pipe; and integral, or welded, attachments. Nonintegral attachments include vendor-supplied hardware such as clamps, U-bolts, pipe straps, clevis hangers, etc., as well as engineered attachments such as saddles and line contact with structural steel. In cases where the attachment is nonintegral and broadly fits the contours of the pipe (as with the above-mentioned vendor hardware), local stresses usually need not be considered, since the load is removed relatively evenly throughout the pipe's half perimeter.

8.3.1.1 Saddles. When the engineer designs a saddle, as shown in Fig. 8.7, the formula for the stresses induced in the pipe wall is given in Roark and Young's *Formulas for Stress and Strain*. This formula states that the maximum stress developed by a saddle is

$$S_{max} = \frac{KP}{t^2} \ln \frac{R}{t} \quad \text{psi (N/mm}^2\text{)} \tag{8.16}$$

where $K = 0.02 - 0.00012(\beta - 90)$
β = angle of contact between saddle and pipe (°)
P = total saddle load, lb (N)

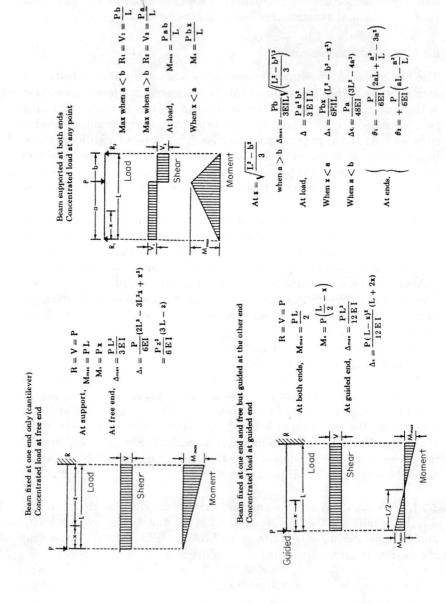

Beam fixed at one end only (cantilever)
Concentrated load at free end

$$R = V = P$$
At support, $M_{max} = PL$
$$M_z = Px$$
At free end, $\Delta_{max} = \frac{PL^3}{3EI}$
$$\Delta_z = \frac{P}{6EI}(2L^3 - 3L^2z + z^3)$$
$$= \frac{Pz^2}{6EI}(3L - z)$$

Beam fixed at one end and free but guided at the other end
Concentrated load at guided end

$$R = V = P$$
At both ends, $M_{max} = \frac{PL}{2}$
$$M_z = P\left(\frac{L}{2} - x\right)$$
At guided end, $\Delta_{max} = \frac{PL^3}{12EI}$
$$\Delta_z = \frac{P(L-x)^2}{12EI}(L + 2x)$$

Beam supported at both ends
Concentrated load at any point

Max when $a < b$ $R_1 = V_1 = \frac{Pb}{L}$

Max when $a > b$ $R_2 = V_2 = \frac{Pa}{L}$

At load, $M_{max} = \frac{Pab}{L}$

When $x < a$ $M_x = \frac{Pbx}{L}$

At $x = \sqrt{\frac{L^2 - b^2}{3}}$

when $a > b$ $\Delta_{max} = \frac{Pb\sqrt{(L^2 - b^2)^3}}{3EIL\sqrt{3}}$

At load, $\Delta = \frac{Pa^2b^2}{3EIL}$

When $x < a$ $\Delta_z = \frac{Pbx}{6EIL}(L^2 - b^2 - x^2)$

When $a < b$ $\Delta_z = \frac{Pa}{48EI}(3L^2 - 4a^2)$

At ends, $\theta_1 = -\frac{P}{6EI}\left(2aL + \frac{a^3}{L} - 3a^2\right)$

$\theta_2 = +\frac{P}{6EI}\left(aL - \frac{a^3}{L}\right)$

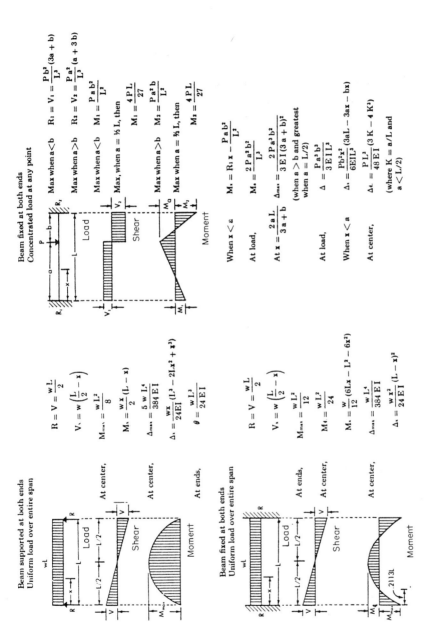

Figure 8.6 Beam formulas. SOURCE: *Design of Welded Structures*, by Omer W. Blodgett. (Courtesy of the James F. Lincoln Arc Welding Foundation.)

t = pipe wall thickness, in (mm)

R = pipe mean radius, in (mm)

The stresses given by Eq. (8.16) are circumferential bending stresses, rather than longitudinal in the pipe. Therefore the resultant stresses should be combined not with the longitudinal pressure and moment stresses, but rather with the circumferential pressure (hoop) stress when the pipe wall stresses are evaluated.

8.3.1.2 Line contact. When the pipe is supported through bearing against structural steel (as in Fig. 8.8), all the force is transmitted through a line contact, theoretically of zero width. This develops additional longitudinal and circumferential stress in the wall.

An accurate model for this case is also presented by Roark and Young. The maximum stresses developed in the pipe wall are given as

$$S_{\text{long}} = -0.15 B^3 P R^{1/4} b^{-1/2} t^{-7/4} \quad \text{psi (N/mm}^2\text{)} \tag{8.17a}$$

$$S_{\text{circ-memb}} = -0.13 B P R^{3/4} b^{-3/2} t^{-5/4} \quad \text{psi (N/mm}^2\text{)} \tag{8.18a}$$

$$S_{\text{circ-bend}} = -B^{-1} P R^{1/4} b^{-1/2} t^{7/4} \quad \text{psi (N/mm}^2\text{)} \tag{8.19a}$$

where $B = [12(1 - \nu^2)]^{1/8}$

ν = Poisson's ratio

P = total applied load, lb (N)

R = pipe mean radius, in (mm)

b = one-half of contact length, in (mm)

t = pipe wall thickness, in (mm)

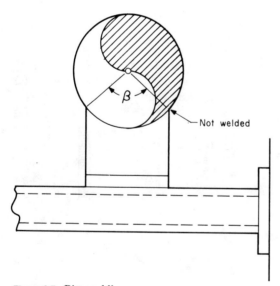

Figure 8.7 Pipe saddle.

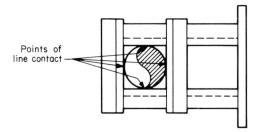

Figure 8.8 Line contact.

Since the engineer is interested in only the absolute value of these stresses, the signs may be dropped in these equations. For steel pipe ($\nu = 0.3$) and given contact length $L = 2b$, these equations reduce to

$$S_{\text{long}} = \frac{0.52 R^{1/4}}{L^{1/2} t^{7/4}} (P) \tag{8.17b}$$

$$S_{\text{circ-memb}} = \frac{1.05 R^{1/4}}{L^{1/2} t^{7/4}} (P) \tag{8.18b}$$

$$S_{\text{circ-bend}} = \frac{0.496 R^{3/4}}{L^{3/2} t^{5/4}} (P) \tag{8.19b}$$

in which units are the same as in Eqs. (8.17a, 8.18a, 8.19a).

The longitudinal stress should be combined with the longitudinal pressure stress and the pipe moment stress to determine the total stress in the pipe wall longitudinally. The two circumferential stresses should be added and combined with the circumferential pressure stress for comparison with the code stress limits.

8.3.1.3 Welded lugs. The two most common types of integral attachments are welded lugs (see Fig. 8.9) and trunnions. Research has been performed on the effects of these attachments on pipe stresses, with the results published by the Welding Research Council (WRC) in Bulletins 107 and 198.

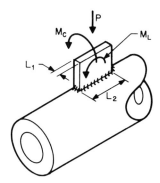

Figure 8.9 Welded lug.

WRC Bulletin 198 deals with welded lugs. The formulas included there for induced stresses due to welded lug attachments are applicable to both shear and bearing lugs, since they allow for longitudinal moment, circumferential moment, and radial load. The effects of shear and torsion are neglected, since they will not cause normal stresses in the pipe wall. The formulas (reprinted here with permission of the Welding Research Council) for the pipe wall stresses induced by rectangular attachments are

$$S_P = C_P \left(\frac{P}{A}\right) \tag{8.20}$$

$$S_L = C_L \left(\frac{M_L}{Z}\right) \tag{8.21}$$

$$S_C = C_C \left(\frac{M_C}{Z}\right) \tag{8.22}$$

where S_P = stress due to radial load, psi (N/mm²)
S_L = stress due to longitudinal moment, psi (N/mm²)
S_C = stress due to circumferential moment, psi (N/mm²)
P = applied radial force, lb (N)
M_L = applied longitudinal moment, in·lb (mm·N)
M_C = applied circumferential moment, in·lb (mm·N)
A = cross-sectional area of pipe, in² (mm²)
Z = section modulus of pipe, in³ (mm³)

$$C_P = 12 \left(\frac{R}{t}\right)^{0.64} \eta^{1.54}$$

$$C_L = 1.2 \left(\frac{R}{t}\right)^{0.74} \eta^{4.74}$$

$$C_C = 1.8 \left(\frac{R}{t}\right)^{0.9} \eta^{3.4}$$

R = pipe mean radius, in (mm)
t = pipe wall thickness, in (mm)

Here η is a value which differs for each type of loading. It is defined as

$$\eta = -(x_1 \cos\theta + y_1 \sin\theta) - \frac{1}{A_0}(x_1 \sin\theta - y_1 \cos\theta)^2$$

where $x_1 = x_0 + \log \beta_1$
$y_1 = y_0 + \log \beta_2$
$\beta_1 = L_1/(2R)$
$\beta_2 = L_2/(2R)$
L_1 = circumferential width of lug, in (mm)
L_2 = longitudinal width of lug, in (mm)

Values for A_0, θ, x_0 and y_0 vary with the load type, as follows:

	A_0	θ	x_0	y_0
P	2.2	40°	0	0.05
M_L	2.0	50°	−0.45	−0.55
M_C	1.8	40°	−0.75	−0.60

Use of the above formulas is restricted to those cases where

$$\left(\frac{\beta_1}{0.3}\right)^2 + \left(\frac{\beta_2}{1.2}\right)^2 \leq 1.0$$

The total stress induced in the pipe wall is equal to the sum of the stresses due to the radial load and the two moments. This induced stress is additive to the longitudinal stresses already present in the pipe.

8.3.1.4 Trunnions. A trunnion is an attachment consisting of a pipe or tube steel welded to the run pipe, as shown in Fig. 8.10. WRC Bulletin 107 outlines the requirements for evaluating welded trunnion attachments.

The formulas for local stresses due to trunnions, like those for lugs, allow for radial force and circumferential and longitudinal moments. Again, shear and torsion may be neglected since these loads do not induce bending in the pipe wall. The applicable formulas (reprinted with the permission of the Welding Research Council) may be rewritten as

$$S_{\text{long}} = \frac{C_1 P}{Rt} + \frac{C_2 6P}{t^2} + \frac{C_3 M_C}{R^2 \beta t} + \frac{C_4 6 M_C}{R \beta t^2} + \frac{C_5 M_L}{R^2 \beta t} + \frac{C_6 M_L}{R \beta t^2} \quad (8.23)$$

$$S_{\text{circ}} = \frac{C_7 P}{Rt} + \frac{C_8 6P}{t^2} + \frac{C_9 M_C}{R^2 \beta t} + \frac{C_{10} 6 M_C}{R \beta t^2} + \frac{C_{11} M_L}{R^2 \beta t} + \frac{C_{12} M_L}{R \beta t^2} \quad (8.24)$$

where S_{long} = total induced longitudinal stress, psi (N/mm²)
S_{circ} = total induced circumferential stress, psi (N/mm²)

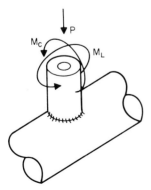

Figure 8.10 Trunnion attachment.

P = applied radial load, lb (N)
M_C = applied circumferential moment, in·lb (mm·N)
M_L = applied longitudinal moment, in·lb (mm·N)
C_1 to C_{12} = experimental constants found on figures published in WRC Bulletin 107 (based on parameters β and γ)
$\beta = 0.875 r_o/R$ (for circular attachment)
$= C/R$ (for square attachment) (dimensionless)
$\gamma = R/t$ (dimensionless)
r_o = outer radius of circular attachment, in (mm)
C = half of width of square attachment, in (mm)
R = mean radius of pipe, in (mm)
t = wall thickness of pipe, in (mm)

The calculated longitudinal stress should be added to the longitudinal pressure stress and the moment stress in the pipe prior to comparison to piping allowable stresses. Note that the piping codes, at present, give no specific local stress limits for qualifying the pressure boundary at the point of attachment. It is left to the individual engineering organization to determine the allowable stress level. The calculated circumferential stress should be added to the circumferential pressure stress for evaluation.

8.3.2 Evaluation of vendors' hardware

This stage of the support evaluation process usually is the simplest.

Standard pipe support hardware has normally been preengineered to the extent that the vendor has on record calculations or test results demonstrating specific load capacities. These capacities are published either as catalog ratings or as a load capacity data sheet (LCDS), as shown in Table 8.2. LCDSs show the hardware name, part number, and load capacities for various sizes and service levels.

Since the evaluation of the hardware capacities has already been done by others, engineering analysis time is reduced to the time necessary to compare the actual load to the allowable load and to select the proper size. The use of preengineered vendor hardware leads to reduced engineering time, standardized designs, simplified procurement, and usually reduced construction costs.

8.3.3 Structural analysis of support steel

When the engineer must provide structural steel in addition to or in place of preengineered hardware, this steel must be analyzed to verify the suitability of the structure for the design load. Several solution methods are

TABLE 8.2 Typical Load Capacity Data Sheet

XYZ COMPANY
PIPE HANGER DIVISION
QUALIFIED PRODUCT　　　　　　　　　　PART NO. C-100
LOAD RATINGS　　　　　　　　　　　　　ITEM: STANDARD PIPE CLAMP

Pipe size, in	Item	Design loading at levels A and B — Maximum load rating, lb at 650°F	Level C — Maximum load rating, lb at 650°F	Level D — Maximum load rating, lb at 650°F
$\frac{1}{2}$	C-100	500	670	940
$\frac{3}{4}$	↑	500	670	940
1"		500	670	940
$1\frac{1}{4}$		500	670	940
$1\frac{1}{2}$		550	730	1030
2		1040	1380	1960
$2\frac{1}{2}$		1040	1380	1960
3		1040	1380	1960
$3\frac{1}{2}$		940	1250	1770
4		940	1250	1770
5		940	1250	1770
6		1615	2150	3040
8		1615	2150	3040
10		2490	3310	4680
12		2490	3310	4680
14		2490	3310	4680
16		2490	3310	4680
18		3060	4070	5750
20		3060	4070	5750
24	↓	3060	4070	5750

available to the engineer, a few of which are discussed in the following pages.

Pipe support analysis may be significantly simplified by optimizing the design of the structure and connections such that the problem may be solved by statics. In problems which are statically determinant, the arrangement of the reactions (determined by the connection details) dictates the distribution of the forces and moments within the structure. For statically indeterminate problems, the analyst must evaluate the relative effects of cross-sectional and material properties of the various members upon load distribution in the structure. Obviously, this is a more time-consuming analysis.

The designer can assure that the structure is determinate by providing an equal number of reactions as there are force and moment equilibrium equations to be solved. For a structure in three-dimensional space, these equilibrium equations may be stated as:

$$\Sigma F_x = 0 \quad \Sigma M_x = 0$$
$$\Sigma F_y = 0 \quad \Sigma M_y = 0$$
$$\Sigma F_z = 0 \quad \Sigma M_z = 0$$

These equations state that the sum of the forces and moments (including pipe loads and reactions) applied to the system must equal zero (if the net load were not zero, the system would not be in equilibrium and would be forced to move). If the number of unknown reactions (i.e., six) is the same as the number of equations, the solution may be found easily. For a two-dimensional structure, the number of equations may be reduced to three—summing the forces along the two in-plane axes and the moment about the out-of-plane axis.

The number of reactions acting on the structure is determined not only by the number of attachment points to the building, but also by the boundary conditions at those points. The two most commonly modeled boundary conditions are the extreme ends of the restraint spectrum. These are fixed (or rigid) and pinned conditions. The fixed boundary prevents rotation and translation of the structure at the point of attachment and therefore creates a moment reaction as well as a force. This connection may be best provided in the field by all-around welding of a member. The pinned end permits member rotation at the connection and therefore transmits only forces as reactions. This connection is best simulated by welding only the webs of wide flanges or by using a clip angle connection as detailed in the AISC Manual of Steel Construction. Both of these details permit free rotation of the flanges. Representations of a fixed connection and a pinned connection are shown in Figs. 8.11 and 8.12, respectively.

When attaching to building steel, which normally consists of wide

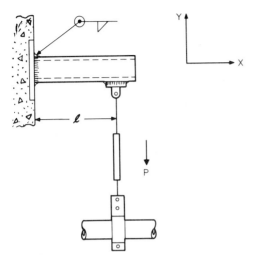

Figure 8.11 Fixed connection between tube steel and embedded plate.

flanges, pinned connections are normally preferred to fixed connections. This is in order to avoid the application of torsional moments to open sections, which are notoriously weak when loaded in that manner.

The structures in Figs. 8.11 and 8.12 are statically determinate. The cantilever in Fig. 8.11 has one connection, which provides a reaction in all 6 degrees of freedom. Four of the equilibrium equations may be dismissed

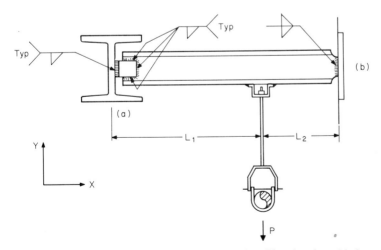

Figure 8.12 Pinned connections: (*a*) clip angles; (*b*) web only welded.

immediately since the structure is loaded only in the F_y and the M_z directions. Therefore

$$\Sigma F_y = 0 \quad R_y - P = 0 \quad R_y = P$$
$$\Sigma M_z = 0 \quad M_z - PL = 0 \quad M_z = PL$$

The reactions at the attachment are the force P along the y axis and the moment Pl about the z axis.

In Fig. 8.12, the structure is also loaded in only 2 degrees of freedom (F_y and M_z), so only two equations need to be solved. Summing the forces and the moments (about point a) gives

$$\Sigma F_y = 0 \quad R_a + R_b - P = 0$$
$$\Sigma M_Z = 0 \quad -PL_1 + R_b(L_1 + L_2) = 0$$
$$R_a = \frac{PL_2}{L_1 + L_2} \quad R_b = \frac{PL_1}{L_1 + L_2}$$

Note that the sum of the moments may be taken about any point, whether located in the frame or arbitrarily selected in space.

When the number of unknown reactions exceeds the number of equilibrium equations, the structure is considered to be statically indeterminate. To solve for the load distribution in these structures, sufficient equations must be added to make the problem solvable. The equations that can be added are equations of compatibility which state that rotations and displacements of the joints must be such that the relation of the members joined at these points must be the same after loading as prior to loading (i.e., fixed at right angles, etc.). The equations of compatibility dictate that the load is distributed through the structure according to the relative stiffness of the members, with the stiffer members accepting the higher load.

An example is shown in Fig. 8.13. A load of 1000 is applied axially at point B. Both reactions points (at A and C) are fully restrained in the direction of loading. For this configuration, only one equilibrium equation ($\Sigma F_{\text{axial}} = 0$) can be written, while there are two unknown F_x reactions. Therefore *one* additional equation of compatibility must be added. We know that point B will displace under load toward point A. Since the

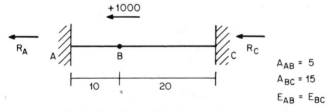

Figure 8.13 Statically indeterminate beam.

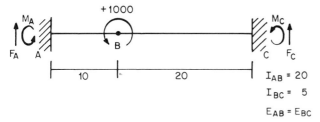

Figure 8.14 Statically indeterminate beam with moment loading.

structure remains continuous, the elongation of member B-C must be equal to the contraction of member A-B. This provides the second equation: $\Delta_{A-B} = -\Delta_{B-C}$.

Axial displacement in a member is equal to $\Delta = PL/(AE)$. Therefore the loading distribution in Fig. 8.13 may be determined by solving the following equations:

$$R_A + R_C + 1000 = 0$$

$$\frac{-R_A(10)}{5E} + \frac{R_C(20)}{15E} = 0$$

Thus,

$$R_A = -400 \qquad R_C = -600$$

Problem 8.2 This method is equally useful when moment loadings are applied. In this case, or when additional compatibility equations are needed, it is useful to equate rotations as well as displacements. As shown in Fig. 8.14, two equilibrium equations ($\Sigma F_{\text{shear}} = 0$ and $\Sigma M = 0$) are available, but there are four unknown reactions (F_A, M_A, F_C, M_C). This means that two compatibility equations must be added.

Two convenient equations to add are those which solve for the rotation and displacement at point C, which are known to be zero. Therefore the four equations are

$$F_A + F_C = 0$$

$$M_A + M_C + 30F_C + 1000 = 0$$

From Eqs. (8.14) and (8.15),

$$\theta_C = \frac{M_A(10)}{E(20)} - \frac{F_A(10)^2}{2E(20)} + \frac{1000(20)}{E(5)} + \frac{[M_A - F_A(10)](20)}{E(5)} - \frac{F_A(20)^2}{2E(5)} = 0$$

$$\Delta_C = \frac{M_A(10)^2}{2E(20)} - \frac{F_A(10)^3}{6E(20)} + \frac{1000(20)^2}{2E(5)} + \frac{[M_A - F_A(10)](20)^2}{2E(5)} - \frac{F_A(20)^3}{6E(5)}$$

$$+ \left[\frac{M_A(10)}{E(20)} - \frac{F_A(10)^2}{2E(20)}\right](20) = 0$$

The latter two equations reduce to

$$4.5M_A - 82.5F_A + 4000 = 0$$

$$52.5M_A - 725F_A + 40{,}000 = 0$$

Now, with four equations and four unknowns, the load distribution may be solved by substitution:

$$F_A = 28.1 \quad M_A = -374.3 \quad F_c = -28.1 \quad M_c = 216.4$$

The results of these two problems demonstrate the principle that a larger part of the load is taken by the stiffer members in the structure. This principle is the basis for the solution method known as *moment distribution*. This method provides a means of distributing applied moment loadings through a frame. The moments distributed are generated from applied loads (fixed-end moments), counteracting moments, and joint rotation. The steps involved in moment distribution are as follows:

1. Determine the relative stiffnesses of all members incident at a joint. This is done by dividing the stiffness (I/l ratio) of the member of interest by the sum of the I/l ratios of all members incident at the joint.

2. Determine the applied loads. This is done by calculating the moments generated at the ends of the loaded member as though the ends were fixed. These moments, known as the *fixed-end moments*, may be found by consulting the beam formulas in Fig. 8.6. A recommended sign convention is to call positive those moments acting clockwise on the ends of the members.

3. The unbalanced moments at the joints must be counteracted by equal moments of the opposite direction. These moments are applied to the members incident at the joint according to the relative stiffnesses of the members.

4. The balancing of the joint throws the moment to the far end of the incident members. This moment is of the same sign and is one-half of the magnitude of the counteracted moment. These rotation-generated moments cause unbalanced moments at other joints, which must be balanced as above. The iteration is continued until the results are within an acceptable tolerance.

5. Through frame and member equilibrium, the reaction and internal member forces may be calculated.

Problem 8.3 An example of moment distribution is shown in Fig. 8.15. The first step is to find the relative stiffnesses of the members. The I/l ratio of members A-B and C-D is $\frac{10}{25} = 0.4$. The I/l ratio of member B-C is $\frac{5}{20} = 0.25$. Therefore the distribution factor for members A-B and C-D is $0.4/(0.4 + 0.25) = 0.615$, and for member B-C it is $0.25/(0.4 + 0.25) = 0.385$.

The fixed-end moments on member B-C are equal to $1000(20)/8$, or 2500. Given the sign convention which calls positive those moments that act clockwise on the end of a member, the fixed-end moments are -2500 at B and $+2500$ at C.

The unbalanced moment (-2500) at B must be balanced by a moment of

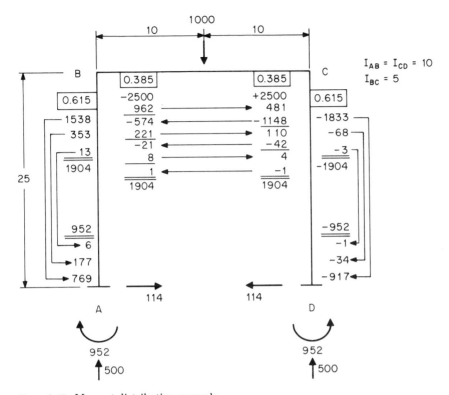

Figure 8.15 Moment distribution example.

+2500. This 2500 is distributed according to the ratios of the stiffnesses. Member A-B receives a moment of 0.615(2500) = 1538, and member B-C receives a moment of 0.385(2500) = 962.

The carryover moment is equal to one-half of the resisting moment and has the same sign. Therefore, a moment of 769 is carried over to A, and 481 is carried to C.

Now joint C must be balanced. The same procedure is followed, with the carryover moment to B causing the joint to become unbalanced. This procedure is repeated until the result converges.

Once the moment reactions have been determined, the forces can be determined by equilibrium. By frame equilibrium, the vertical force reactions are found to be +500. The horizontal reactions can be found by taking the equilibrium of members A-B and C-D. The member shear is found by dividing the sum of the end moments by the member length, or $V = (952 + 1904)/25 = 114$.

From these forces and moments, or those calculated by other methods, member stresses can be calculated. The allowable stresses for the steel used in pipe supports are regulated, for non-safety-related piping, by the design specification of the AISC. The design rules for the structural steel of nuclear pipe supports are governed by the rules of Appendix XVII of

Section III of the ASME Boiler and Pressure Vessel Code. The stress requirements of Appendix XVII closely follow those of the AISC code, except that they permit increased stress allowables for occasional loads classified as service level C or D.

8.3.4 Evaluation of structural attachments

8.3.4.1 Evaluation of weldments. Once it has been determined that the stresses developed in the structural members making up the pipe support meet the allowable stresses as dictated by the applicable code, it remains to design the welds forming the joints between member and pipe, member and member, and member and building structure.

The capacity of a weld is a function of the weld pattern properties, the size of the weld throat, and the types of electrode and base metal used in the weld. To evaluate a weld, determine the forces and moments acting at the joint; apply these loads to the weld pattern properties to find the force on the weld per unit length; then determine the weld capacity based on the weld size and materials used. If the capacity of the weld is greater than the force on the weld, the connection is adequate.

The calculation of forces on welds is analogous to the calculation of stresses in members, in that the weld pattern has an area, a section modulus, and a polar moment of inertia. The *area* of the weld A_w is equal to the length of the weld lines in the pattern. The *section modulus* of the weld S_w is equal to the second moment of the weld length about the neutral axis of the weld pattern divided by the distance to the extreme fiber of the weld group. As for member cross sections, there may be a different S_w for each axis of the weld group, as well as for the top and bottom portions about each axis (for nonsymmetric sections). The *polar moment of inertia* J_w is equal to the sum of the second moment of the weld length about two mutually perpendicular neutral axes of the weld group. See Table 8.3 for formulas for S_w and J_w for several weld groups.

The forces on the weld will be in three directions—axial and two shears, along the x and y axes of the weld group. The axial forces are developed by the applied axial force and the bending moments about the weak and strong axes. The shear forces are developed by the applied shears and the torsional moment. The forces in the weld are calculated as follows:

$$f_1 = \frac{F_{\text{axial}}}{A_w} + \frac{M_{Bx}}{S_{wx}} + \frac{M_{By}}{S_{wy}} \tag{8.25}$$

$$f_2 = \frac{F_x}{A_w} + \frac{M_T(\bar{y})}{J_w} \tag{8.26}$$

$$f_3 = \frac{F_y}{A_w} + \frac{M_T(\bar{x})}{J_w} \tag{8.27}$$

where f_1, f_2, f_3 = mutually perpendicular forces per unit length developed in weld, lb/in (N/mm)

A_w = area of weld, in (mm)

S_{wx}, S_{wy} = section modulus of weld about the x and y axes, respectively, in^2 (mm^2)

J_w = polar moment of inertia of weld, in^3 (mm^3)

F_{axial} = axial (perpendicular to plane of weld) force on weld, lb (N)

M_{Bx}, M_{By} = bending moment about the x and y axes, respectively, in·lb (mm·N)

F_x, F_y = shear forces along the x and y axes, respectively, lb (N)

M_T = torsional moment on weld, in·lb (mm·N)

$\bar{x}, \bar{y}$ = distance to point of interest from neutral axis, along the x and y axes, respectively, in (mm)

The resultant force on the weld f_w is found by taking the square root of the sum of the squares of the three perpendicular forces acting on the weld:

$$f_w = \sqrt{f_1^2 + f_2^2 + f_3^2} \quad \text{lb/in (N/mm)} \tag{8.28}$$

The capacity of a weld is determined by multiplying the weld throat thickness by the allowable shear stress of the weld material, which is 0.3 times the nominal tensile strength of the weld material. The throat of the weld is the narrowest part of the weld over which the forces must pass. For example, for a penetration weld, the throat thickness is the depth of the penetration; for a fillet weld, the throat thickness is 70.7 percent of the fillet leg size; for a flare bevel weld, the throat thickness is accepted as $\frac{5}{16}R$, where R is the radius of curvature of the welded member. These types of welds are shown in Fig. 8.16; the symbols for these welds, and others, are shown in Fig. 8.17.

Therefore the capacity of a 0.25-in (6.4-mm) fillet weld of E70 electrode [E70 indicates 70 kips/in^2 (483 N/mm^2) minimum tensile strength] is found by

$$F_w = t_e(0.3T_u) = 0.707(0.25)(0.3)(70{,}000) = 3712 \text{ lb/in}$$
$$= 0.707(6.4)(0.3)(483) = \text{N/mm}$$

This capacity must exceed the force on the weld f_w calculated above.

8.3.4.2 Evaluation of baseplates. The support may be attached to the building structure through welds either to building steel or to baseplates, which are in turn attached to building concrete. These baseplates may be attached to the concrete by two methods—with embedded attachments or with anchor bolts.

TABLE 8.3 Properties of Weld Groups

Outline of welded joint b = width $\quad d$ = depth	Bending (about horizontal axis x-x), in^2	Twisting (about center of gravity), in^3
(vertical line, depth d)	$S_w = \dfrac{d^2}{6}$	$J_w = \dfrac{d^3}{12}$
(two vertical lines, width b, depth d)	$S_w = \dfrac{d^2}{3}$	$J_w = \dfrac{d(3b^2 + d^2)}{6}$
(rectangle with x-axis, width b, depth d)	$S_w = bd$	$J_w = \dfrac{b^3 + 3bd^2}{6}$
$N_y = \dfrac{b^2}{2(b+d)}$, $\;N_x = \dfrac{d}{2(b+d)}$ (L-shape)	$S_w = \dfrac{4bd + d^2}{6} = \dfrac{d^2(4b+d)}{6(2b+d)}$ top $\qquad$ bottom	$J_w = \dfrac{(b+d)^4 - 6b^2d^2}{12(b+d)}$
$N_y = \dfrac{b^2}{2b+d}$ (channel shape)	$S_w = bd + \dfrac{d^2}{6}$	$J_w = \dfrac{(2b+d)^3}{12} - \dfrac{b^2(b+d)^2}{2b+d}$
$N_x = \dfrac{d^2}{b+2d}$ (channel shape)	$S_w = \dfrac{2bd + d^2}{3} = \dfrac{d^2(2b+d)}{3(b+d)}$ top $\qquad$ bottom	$J_w = \dfrac{(b+2d)^3}{12} - \dfrac{d^2(b+d)^2}{b+2d}$

Shape	S_w	J_w
Box, N_y (not given)	$S_w = bd + \dfrac{d^2}{3}$	$J_w = \dfrac{(b+d)^3}{6}$
T-shape, $N_y = \dfrac{d^2}{b+2d}$	$S_w = \dfrac{2bd+d^2}{3} = \dfrac{d^2(2b+d)}{3(b+d)}$ top bottom	$J_w = \dfrac{(b+2d)^3}{12} - \dfrac{d^2(b+d)^2}{b+2d}$
I-shape, $N_y = \dfrac{d^2}{2(b+d)}$	$S_w = \dfrac{4bd+d^2}{3} = \dfrac{4bd^2+d^3}{6b+3d}$ top bottom	$J_w = \dfrac{d^3(4b+d)}{6(b+d)} + \dfrac{b^3}{6}$
H-shape	$S_w = bd + \dfrac{d^2}{3}$	$J_w = \dfrac{b^3+3bd^2+d^3}{6}$
Double-flange	$S_w = 2bd + \dfrac{d^2}{3}$	$J_w = \dfrac{2b^3+6bd^2+d^3}{6}$
Circle	$S_w = \dfrac{\pi d^2}{4}$	$J_w = \dfrac{\pi d^3}{4}$

SOURCE: *Design of Welded Structures*, by Omer W. Blodgett. (Courtesy of the James F. Lincoln Arc Welding Foundation.)

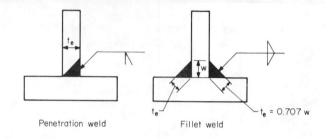

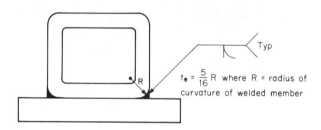

Figure 8.16 Weld effective throat thickness t_e.

The first type, the embedded plate, has attachments such as studs, bent rebar, or structural members welded to it. The plate is then set in place prior to pouring the concrete. The bond between the concrete and the embedded attachment prevents the plate from pulling away from the wall or floor.

Since the available types of embedded attachments vary greatly, it is not possible to provide a general method for estimating embedded plate capacity. In most cases, embedded plates are considered part of the building and so are within the scope of the civil engineer, not the pipe support engineer. The civil engineer should normally provide capacities for the embedded plates, along with an interaction formula for combined loadings. The interaction formula will typically be of the form

$$\left(\frac{f_{\text{axial}}}{F_{\text{axial}}} + \frac{m_x}{M_x} + \frac{m_y}{M_y}\right)^a + \left(\frac{f_x}{F_x} + \frac{f_y}{F_y} + \frac{m_T}{M_T}\right)^b \leq 1.0 \qquad (8.29)$$

where f_{axial} = axial force on embedded plate, lb (N)

f_x, f_y = shear forces on embedded plate along x and y axes, respectively, lb (N)

m_x, m_y = bending moments on embedded plate about x and y axes, respectively, in·lb (mm·N)

m_T = torsional moment acting on embedded plate, in·lb (mm·N)

$F_{\text{axial}}, F_x, F_y, M_x, M_y, M_T$ = capacities of embedded plate for axial force, shear force along the x axis, shear force along the y axis, bending about the x axis, bending about the y axis, and torsion, respectively, lb or in·lb (N or mm·N)

a, b = exponential constants, usually between 1.0 and 2.0 (exact value dependent on plate anchorage configuration)

The second type of baseplate, that using anchor bolts, is not considered part of the building and therefore is within the scope of the support engineer. In order to install this type of baseplate, holes are drilled in the concrete wall or floor, concrete anchors (either of the shell or wedge type) are inserted into the holes, and the plates are bolted to the anchors, against the concrete.

The support engineer must be able to calculate the forces in the anchor bolts and the stresses in the plate in order to evaluate this type of attachment. Once the forces in the anchor bolts are known, the bolts may be evaluated by the following formula:

$$\left(\frac{\text{SF} \times F_T}{\text{RF} \times T_u}\right)^a + \left(\frac{\text{SF} \times F_v}{V_u}\right)^b \leq 1.0 \qquad (8.30)$$

where F_T = tension on anchor bolt, lb (N)

F_v = shear on anchor bolt lb (N)

SF = safety factor for anchor bolt (for nuclear safety applications, this must be 4 for wedge-type anchors and 5 for shell-type anchors)

RF = tension reduction factor, summed for all interfering bolts, due to concrete shear cone interference, based on bolt spacing and embedment

$$= 1 - \sum \left[\frac{2 \cos^{-1}(n_i/l)}{360} - \frac{n_i \sqrt{l^2 - n_i^2}}{\pi l^2}\right]$$

(For calculated values see Table 8.4; also $\cos^{-1}$ should be calculated in degrees.)

n_i = half of the distance to each interfering bolt, in (mm)

l = bolt embedment, in (mm)

T_u = bolt ultimate tensile capacity, lb (N)

V_u = bolt ultimate shear capacity, lb (N)

a, b = exponential constants between 1.0 and 2.0

The method used to calculate bolt loads on a baseplate depends on whether the baseplate is assumed to be rigid or flexible. With a rigid baseplate, the bolt loads developed will be only those necessary to counteract the applied forces and moments. With a flexible baseplate, however, the

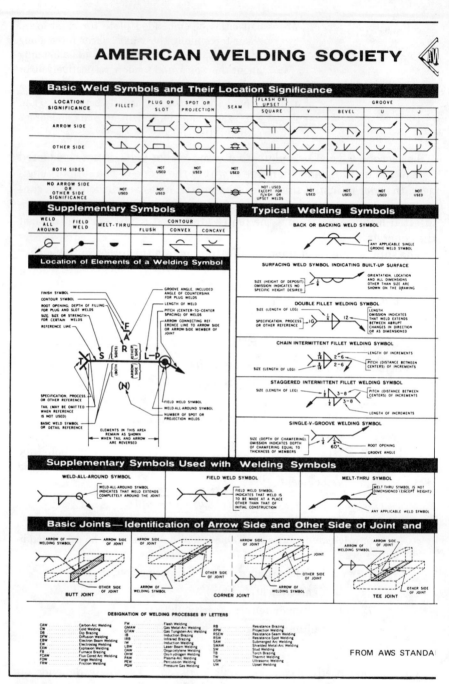

Figure 8.17 Standard welding symbols. *(Reproduced by permission of the American Welding Society.)*

STANDARD WELDING SYMBOLS

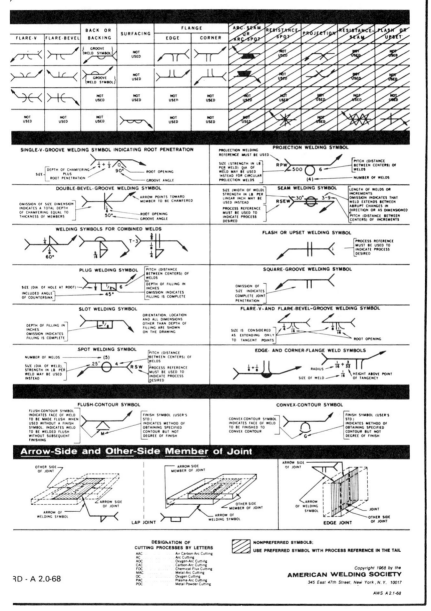

TABLE 8.4 Reduction Factors for Partial Shear Cones—Studs and Inserts

Both sides of shear cone not developed

Case 1

Case 2

Interior studs or inserts in a group. $m < 2 l_e$

m, in	$l_e = 2.5$ in	$l_e = 4$ in	$l_e = 6$ in	$l_e = 8$ in	$l_e = 9$ in	$l_e = 12$ in	$l_e = 15$ in	$l_e = 18$ in
3	0.72	0.47	0.31	0.24	0.21	0.16	0.13	0.11
4	0.90	0.61	0.42	0.31	0.28	0.21	0.17	0.14
5	1.0	0.74	0.51	0.39	0.35	0.26	0.21	0.18
6	1.0	0.86	0.61	0.47	0.42	0.31	0.25	0.21
7	1.0	0.95	0.70	0.54	0.48	0.37	0.29	0.25
8	1.0	1.0	0.78	0.61	0.55	0.42	0.34	0.28
9	1.0	1.0	0.86	0.68	0.61	0.47	0.38	0.31
10	1.0	1.0	0.92	0.74	0.67	0.51	0.42	0.35
11	1.0	1.0	0.97	0.80	0.73	0.56	0.46	0.38
12	1.0	1.0	1.0	0.86	0.78	0.61	0.50	0.42
13	1.0	1.0	1.0	0.91	3.83	0.65	0.53	0.45
14	1.0	1.0	1.0	0.95	0.88	0.70	0.57	0.48
15	1.0	1.0	1.0	0.98	0.92	0.74	0.61	0.51
16	1.0	1.0	1.0	1.0	0.96	0.78	0.65	0.55
17	1.0	1.0	1.0	1.0	0.98	0.82	0.68	0.58
18	1.0	1.0	1.0	1.0	1.0	0.86	0.72	0.61

One side of shear cone not developed

Case 1

Case 2

Exterior studs or inserts in a group
$n < l_e$

n, in	$l_e = 2.5$ in	$l_e = 4$ in	$l_e = 6$ in	$l_e = 8$ in	$l_e = 9$ in	$l_e = 12$ in	$l_e = 15$ in	$l_e = 18$ in
1.5	0.85	0.73	0.66	0.62	0.60	0.58	0.56	0.55
2	0.95	0.80	0.71	0.66	0.64	0.61	0.58	0.57
2.5	1.0	0.86	0.75	0.70	0.67	0.63	0.60	0.59
3	1.0	0.93	0.80	0.73	0.71	0.66	0.63	0.61
4	1.0	1.0	0.89	0.80	0.77	0.71	0.67	0.64
5	1.0	1.0	0.96	0.87	0.83	0.76	0.71	0.67
6	1.0	1.0	1.0	0.93	0.89	0.80	0.75	0.71
7	1.0	1.0	1.0	0.97	0.94	0.85	0.79	0.74
8	1.0	1.0	1.0	1.0	0.98	0.89	0.82	0.77
9	1.0	1.0	1.0	1.0	1.0	0.93	0.86	0.80
10	1.0	1.0	1.0	1.0	1.0	0.96	0.89	0.83
11	1.0	1.0	1.0	1.0	1.0	0.99	0.92	0.86
12	1.0	1.0	1.0	1.0	1.0	1.0	0.95	0.89
13	1.0	1.0	1.0	1.0	1.0	1.0	0.97	0.92
14	1.0	1.0	1.0	1.0	1.0	1.0	0.99	0.94
15	1.0	1.0	1.0	1.0	1.0	1.0	1.0	0.96
16	1.0	1.0	1.0	1.0	1.0	1.0	1.0	0.98
17	1.0	1.0	1.0	1.0	1.0	1.0	1.0	0.99
18	1.0	1.0	1.0	1.0	1.0	1.0	1.0	1.0

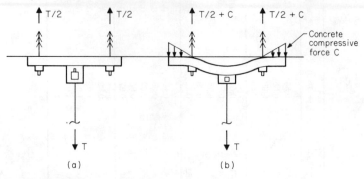

Figure 8.18 Rigid versus flexible baseplates: (*a*) rigid; (*b*) flexible.

tension forces may increase due to prying—the action of the baseplate bending against the concrete. This is illustrated in Fig. 8.18. Actually, all baseplates have a certain degree of flexibility and will develop prying forces. As the rigidity of the baseplate increases (with the ratio of the plate thickness to the length, or as stiffeners are added), the prying forces approach zero. The engineer should analyze all baseplates as though they were flexible until sufficient experience has been gained to permit neglecting prying forces.

One method of analyzing rigid baseplates is to model their interaction with the concrete as one would a reinforced concrete beam. The assumption, as shown in Fig. 8.19, is that an applied moment is resisted by tension in the bolt and compression in the concrete. Since the baseplate is rigid, the strain in the bolt and the concrete must be proportional to their distance from the neutral axis, or pivot point. Therefore, the force distribution and pivot point location will depend on the relative areas and moduli of elasticity of the bolts and the concrete bearing area. Based on

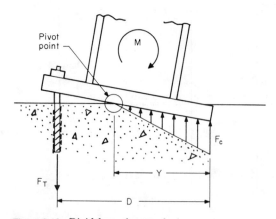

Figure 8.19 Rigid-baseplate analysis.

these assumptions, the distance from the neutral axis to the edge of the plate (Y) can be calculated as

$$Y = \frac{\sqrt{A_B^2(E_s/E_c)^2 + 2A_B(E_s/E_c)DW} - A_B(E_s/E_c)}{W} \quad \text{in (mm)} \quad (8.31)$$

where A_B = total area of bolts in tension, in² (mm²)
E_s = modulus of elasticity of steel bolt, kips/in² (kN/mm²)
E_c = modulus of elasticity of concrete, kips/in² (kN/mm²)
D = distance from bolts in tension to compression edge of plate, in (mm)
W = width of plate, in (mm)

Once the location of the neutral axis is known, the moment arm of the concrete-bolt couple is the distance between the bolt and the resultant of the triangular concrete stress distribution, or $D - Y/3$. Therefore, the tensile force on each bolt F_T and the maximum compressive stress on the concrete σ_c may be calculated as

$$F_T = \frac{M}{n(D - Y/3)} \quad \text{lb (N)} \quad (8.32)$$

$$\sigma_c = \frac{2M}{(D - Y/3)WY} \quad \text{psi (N/mm}^2\text{)} \quad (8.33)$$

where M = applied moment and n = number of bolts in tension.

Applied forces are distributed between the bolts in tension and the concrete in compression, inversely proportional to the distance from the force application to each of these points. A tensile force on the plate will increase the bolt tension and decrease the concrete compression; a compressive force will reduce the bolt tension and increase the concrete compression.

The prying effects of flexible baseplates do not lend themselves easily to manual calculation and are most accurately simulated with computerized finite-element analyses. In lieu of more accurate or easily applied calculation methods, the reader is referred to the AISC formula for estimating prying in steel-to-steel connections using high-strength bolts. Use of the following formula, adapted from the formula for A325 bolts, should give conservative results when it is applied to baseplates using concrete anchors:

$$F_p = F_T \left(1 + \frac{100bd^2 - 18wt^2/n}{70ad^2 + 21wt^2/n}\right) \quad (8.34)$$

where F_P = tensile force in bolt, including prying, lb (N)
F_T = calculated tensile force in bolt without regard to prying, lb (N)
b = distance from edge of structural attachment to line of bolts in tension, in (mm)
d = bolt diameter, in (mm)

t = thickness of plate, in (mm)

a = distance from line of bolts in tension to tension edge of plate, not to exceed $2t$, in (mm)

W = width of plate, in (mm)

See Fig. 8.20. Once the tension in the bolts has been calculated, the stress in the baseplate may be conservatively estimated as

$$\sigma_{Pl} = \frac{6nF_p b}{wt^2} \quad \text{psi (N/mm}^2\text{)} \tag{8.35}$$

Shear forces applied to a baseplate are normally considered to be distributed equally among all the anchor bolts. Shear forces caused by applied torsional moments are distributed according to

$$F_x = \frac{M_T \bar{y}}{J_B} \tag{8.36a}$$

and

$$F_y = \frac{M_T \bar{x}}{J_B} \tag{8.36b}$$

where F_x, F_y = shear force on each bolt along the x and y axes, respectively, lb (N)

M_T = applied torsional moment, in·lb (mm·N)

$\bar{x}$, $\bar{y}$ = distance to the bolt of interest from x and y neutral axes, respectively, in (mm)

J_B = polar moment of inertia of the bolt pattern, found by summing the second moment of the bolts about the x and y neutral axes

= $\Sigma \bar{x}_i^2 + \Sigma \bar{y}_i^2$, in^2 (mm^2)

The x and y shear forces found above would be added to the x and y shear forces due to applied forces and then combined to find the resultant shear F_v:

$$F_v = \sqrt{F_x^2 + F_y^2} \quad \text{lb (N)} \tag{8.37}$$

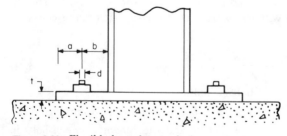

Figure 8.20 Flexible-baseplate analysis.

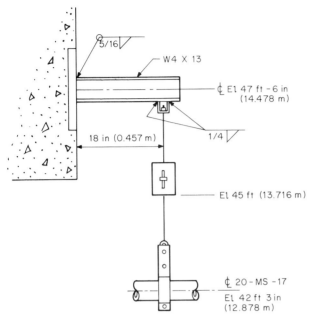

Figure 8.21 Problem 8.4.

This resultant shear force is then used with the calculated tension in the anchor bolt interaction formula to check the anchor bolt capacity.

8.4 Sample Problems

The following sample problems are provided to illustrate the use of the methods explained in this chapter.

Problem 8.4 The engineer is asked to select vendor's hardware and check the conceptual design shown in Fig. 8.21. This figure shows a pipe with a spring assembly attached to a cantilever, which is in turn welded to an embedded plate on the concrete wall. The following information is given:

Piping code:	B31.1
Pipe diameter:	20 in (500 mm)
Design temperature:	600°F (316°C)
Deadweight of pipe:	10,000 lb (44,500 N)
Thermal movement of pipe:	0.35 in (8.9 mm) up

The cantilever is W4 × 13; from AISC member property tables,

$d = 4.16$ in (105.7 mm)

$t_w = 0.28$ in (7.1 mm)

$b = 4.06$ in (103.1 mm)

$S = 5.45$ in^3 (8.93 × 10^4 mm^3)

The embedded plate (capacities and interaction formula are provided by the civil engineering department) has these capacities:

$F_\text{axial} = 20{,}000$ lb (89,000 N)

$F_x, F_y = 25{,}000$ lb (111,250 N)

$M_x, M_y = 260{,}000$ in·lb (2.94 × 10^7 mm·N)

$M_T = 350{,}000$ in·lb (3.96 × 10^7 mm·N)

The interaction formula is

$$\left(\frac{f_\text{axial}}{F_\text{axial}} + \frac{m_x}{M_x} + \frac{m_y}{M_y}\right)^{5/3} + \left(\frac{f_x}{F_x} + \frac{f_y}{F_y} + \frac{m_T}{M_T}\right)^{5/3} \le 1.0$$

Since a clamp is used, local pipe wall stress need not be checked (the pipe stress at this point must be verified by the stress analyst, of course, but no additional stress is assumed to be developed by the pipe attachment in this case).

The clamp, weldless eye nut, rods, welded beam attachment, and spring are all vendor's hardware and must be selected from load capacity data sheets (LCDSs). Based on a preliminary load of 10,000 lb (44,500 N), the following hardware may be chosen from the LCDSs in Fig. 8.22a to d:

Part	Capacity
20-in (500-mm) diameter heavy three-bolt pipe clamp	15,000 lb (66,750 N) @ 600°F (316°C)
Weldless eye nut, size 12	11,630 lb (51,754 N)
1.5-in (37.5-mm) diameter threaded rod	11,630 lb (51,754 N)
Welded beam attachment, size 12	11,630 lb (51,754 N)

The spring must be selected according to both the pipe weight and the expected thermal movement, by using a table such as Table 6.2. The hot load (HL) of the pipe is found by adding the pipe weight to the weight of the hardware supported by the spring:

HL = 10,000 + 125 + 6 + 1.25 ft × 6 lb/ft = 10,139 lb
           (clamp)   (eye nut)   (rod)

HL = 44,500 + 556 + 27 + 0.381 m × 87.6 N/m = 45,116 N

From Table 6.2, we see that the calculated hot load is within the range of a size 17 spring. If we try a short-range spring, with a spring rate of 4000 lb/in (701 N/mm), the cold load (CL) is

CL = 10,139 + 0.35(4000) = 11,539 lb

CL = 45,116 + 8.9(701) = 51,355 N

Referring again to Table 6.2, we note that the calculated cold load is still within the spring range. Checking the variability yields

$$V = \frac{11{,}539 - 10{,}139}{10{,}139} = \frac{51{,}355 - 45{,}116}{45{,}116} = 0.14 \le 0.25$$

Therefore, the size 17 short-range spring is acceptable. Since the cold load is higher than the original pipe load, the vendor's hardware must be rechecked for this higher load. In this case, all the hardware is suitable for the cold load.

Next, check the weld between the beam attachment and the W4 × 13 cantilever. The width of the beam attachment is 7 in (177.8 mm). However, since the beam width is only 4.06 in (103.1 mm), the area of the weld is

$$A_w = 2(4.06) = 8.12 \text{ in } (206.2 \text{ mm})$$

The force on the weld is

$$f_w = \frac{F}{A_w} = \frac{11,539}{8.12} = 1421 \text{ lb/in } (249 \text{ N/mm})$$

The $\frac{1}{4}$-in (6.35-mm) fillet weld has a capacity (assuming E70 electrode) of

$$\begin{aligned} F_w &= t_e \times 0.3 \times T_u \\ &= 0.707 \times 0.25 \times 0.3 \times 70,000 = 3712 \text{ lb/in} \geq 1421 \text{ lb/in} \\ &= 0.707 \times 6.35 \times 0.3 \times 483 = 656 \text{ N/mm} \geq 240 \text{ N/mm} \end{aligned}$$

Therefore this weld is adequate.

Checking the W4 × 13 cantilever for bending reveals

$$\begin{aligned} \sigma_B &= \frac{PL}{S} \\ &= \frac{(11,539)(18)}{5.45} = 38,110 \text{ psi} \\ &= \frac{(51,355)(457)}{8.93 \times 10^4} = 263 \text{ N/mm}^2 \end{aligned}$$

This stress exceeds the allowable bending stress (per AISC code), which is $0.66S_y$ for compact sections or 23,760 psi (164 N/mm²) for 36-kips/in² (248-N/mm²) steel. The W4 × 13 is not good for this application; therefore try a W6 × 20. From member property tables,

$$d = 6.2 \text{ in } (157.5 \text{ mm}) \qquad b = 6.018 \text{ in } (152.9 \text{ mm})$$
$$t_w = 0.258 \text{ in } (6.55 \text{ mm}) \qquad S = 13.4 \text{ in}^2 \ (2.2 \times 10^5 \text{ mm}^3)$$

In this application, the W6 × 20 meets the requirements of a compact section. Therefore the allowable stresses are 23,760 psi (164 N/mm²) for bending and 0.4 (36.000) = 14,400 psi (99 N/mm²) for shear across the web. Thus

$$\begin{aligned} \sigma_B &= \frac{(11,539)(18)}{13.4} = 15,500 \text{ psi} \leq 23,760 \text{ psi} \\ &= \frac{(51,355)(457)}{2.2 \times 10^5} = 107 \text{ N/mm}^2 \leq 164 \text{ N/mm}^2 \end{aligned}$$

$$\begin{aligned} \tau = \frac{V}{dt_w} &= \frac{11,539}{6.2(0.258)} = 7214 \text{ psi} \leq 14,400 \text{ psi} \\ &= \frac{51,355}{157.5(6.55)} = 50 \text{ N/mm}^2 \leq 99 \text{ N/mm}^2 \end{aligned}$$

Therefore the W6 × 20 is acceptable and should be used in place of the W4 × 13.

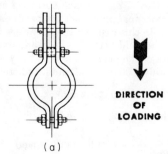

TEMPERATURE 650 °F

PIPE SIZE	DESIGN CONDITION	OPERATING CONDITIONS		
		LEVEL A & B NORMAL & UPSET	LEVEL C EMERGENCY	LEVEL D FAULTED
3/4*	1500	1500	2000	2800
1 *	1500	1500	2000	2800
1-1/4	1500	1500	2000	2800
5	5000	5000	6600	9400
6 *	8000	8000	10600	15000
8 *	8000	8000	10600	15000
10	11000	11000	14600	20600
12	11000	11000	14600	20600
14	12700	12700	16900	23800
16	12700	12700	16900	23800
18	12700	12700	16900	23800
20 **	15000	15000	20000	28200
24 **	15000	15000	20000	28200
28 **	15000	15000	20000	28200
30 **	15000	15000	20000	28200
36 **	15000	15000	20000	28200

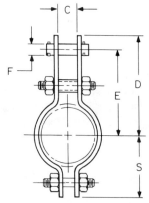

PIPE SIZE	C	D	E	F	S	STOCK SIZE	WEIGHT
20	$2\frac{1}{4}$	21	18	2	$16\frac{1}{4}$	1 x 5	125

Figure 8.22 Load capacity data sheet: (a) heavy three-bolt pipe clamp. *(Courtesy of Bergen-Paterson Pipesupport Corp.)*

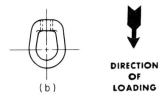

(b)

DIRECTION OF LOADING

TEMPERATURE 350 °

SIZE	DESIGN CONDITION	OPERATING CONDITIONS		
		LEVEL A & B NORMAL & UPSET	LEVEL C EMERGENCY	LEVEL D FAULTED
4	1130	1130	1500	2260
5	1810	1810	2400	3620
6	2710	2710	3600	5420
7	3770	3770	5000	7540
8	4960	4960	6600	9920
10	8000	8000	10600	16000
12	11630	11630	15500	23260
14	15700	15700	20900	31400
16	20700	20700	27600	41400
18	27200	27200	36200	54400
20	33500	33500	44600	67000

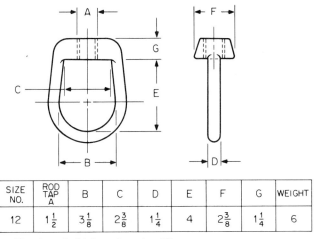

SIZE NO.	ROD TAP A	B	C	D	E	F	G	WEIGHT
12	$1\frac{1}{2}$	$3\frac{1}{8}$	$2\frac{3}{8}$	$1\frac{1}{4}$	4	$2\frac{3}{8}$	$1\frac{1}{4}$	6

Figure 8.22 (*Continued*) (*b*) forged steel weldless eye nut.

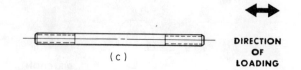

(c)

DIRECTION OF LOADING

TEMPERATURE 350

SIZE	DESIGN CONDITION	OPERATING CONDITIONS		
		LEVEL A & B NORMAL & UPSET	LEVEL C EMERGENCY	LEVEL D FAULTED
1/2	1130	1130	1500	2120
5/8	1810	1810	2400	3400
3/4	2710	2710	3600	5090
7/8	3770	3770	5000	7080
1	4960	4960	6600	9320
1-1/4	8000	8000	10600	15040
1-1/2	11630	11630	15500	21860
1-3/4	15700	15700	20900	29510
2	20700	20700	27600	38910
2-1/4	27200	27200	36200	51130
2-1/2	33500	33500	44600	62980

Figure 8.22 (*Continued*) (*c*) machine thread rod.

Next the weld to the embedded plate must be checked. From Table 8.3, the properties of the all-around weld are

$$A_w = 4(6.018) + 2(6.2) = 36.5 \text{ in } (926 \text{ mm})$$

$$S_w = 2(6.018 \times 6.2) + \frac{(6.2)^2}{3} = 87.4 \text{ in}^2 \text{ } (5.64 \times 10^4 \text{ mm}^2)$$

The force on the weld is

$$f_w = \sqrt{\left(11{,}539 \times \frac{18}{87.4}\right)^2 + \left(\frac{11{,}539}{36.5}\right)^2} = 2397 \text{ lb/in}$$

$$= \sqrt{\left(51{,}355 \times \frac{457}{5.64 \times 10^4}\right)^2 + \left(\frac{51{,}355}{926}\right)^2} = 420 \text{ N/mm}$$

The weld capacity of the $\frac{5}{16}$-in (7.94-mm) fillet weld is

$$F_w = 0.707 \times 0.3125 \times 0.3 \times 70{,}000 = 4640 \text{ lb/in} \geq 2397 \text{ lb/in}$$

$$= 0.707 \times 7.94 \times 0.3 \times 483 = 813 \text{ N/mm} \geq 420 \text{ N/mm}$$

Therefore the weld is adequate. Using the interaction formula to check the embedded plate shows

$$\left(\frac{0}{20{,}000} + \frac{11{,}539(18)}{260{,}000} + \frac{0}{260{,}000}\right)^{5/3} + \left(\frac{0}{25{,}000} + \frac{11{,}539}{25{,}000} + \frac{0}{350{,}000}\right)^{5/3} = 0.96 \leq 1.0$$

$$\left(\frac{0}{89{,}000} + \frac{51{,}355(457)}{2.94 \times 10^7} + \frac{0}{2.94 \times 10^7}\right)^{5/3} + \left(\frac{0}{111{,}250} + \frac{51{,}355}{111{,}250} + \frac{0}{3.96 \times 10^7}\right)^{5/3} = 0.96 \leq 1.0$$

The embedded plate is adequate.

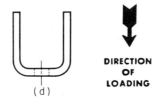

(d)

DIRECTION OF LOADING

TEMPERATURE 350

SIZE	DESIGN CONDITION	OPERATING CONDITIONS		
		LEVEL A & B NORMAL & UPSET	LEVEL C EMERGENCY	LEVEL D FAULTED
4	1130	1130	1500	2120
5	1810	1810	2400	3400
6	2710	2710	3600	5090
7	3770	3770	5000	7080
8	4960	4960	6600	9320
10	8000	8000	10600	15040
12	11630	11630	15500	21860

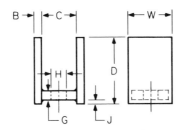

SIZE	B	C	D	G	H HOLE	J	W
12	$\frac{1}{2}$	6	$7\frac{1}{2}$	$1\frac{1}{4}$	$1\frac{5}{8}$	$\frac{5}{16}$	7

Figure 8.22 (*Continued*) (*d*) adjustable welded beam attachment.

Problem 8.5 Figure 8.23a shows a snubber running between a pipe (attached by a trunnion) and a knee brace. The knee-brace members are mounted to the floor by baseplates held with anchor bolts. The following information is given:

Piping code:	Nuclear Class 3
Pipe size:	12-in (300-mm) nominal diameter, standard schedule pipe $D = 12.75$ in (324 mm), $t = 0.375$ in (9.525 mm)
Pipe material allowable stress:	A 106 GrB carbon steel, $S_h = 15{,}000$ psi (103 N/mm^2)
Existing pipe stresses:	8240 psi (57 N/mm^2), upset; 12,580 psi (87 N/mm^2), emergency; circumferential pressure stress $= 4800$ psi (33.2 N/mm^2)

260 Piping and Pipe Support Systems

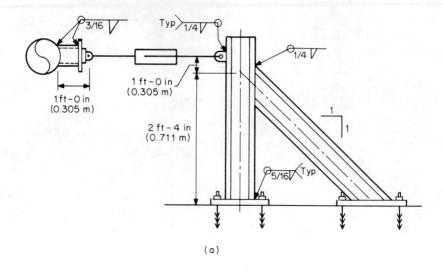

(a)

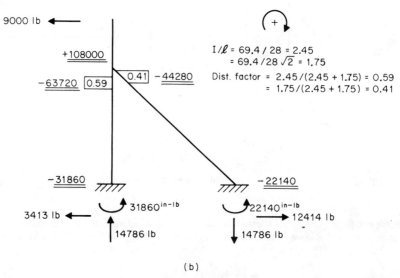

(b)

Figure 8.23 Problem 8.5.

Trunnion size: 6-in (150-mm) nominal diameter, standard schedule pipe, $D = 6.625$ in (168.3 mm), $t = 0.28$ in (7.11 mm), $r = 2.245$ in (57.02 mm), $A = 5.58$ in^2 (3600 mm^2)

Thermal movement: 2 in (50.8 mm) laterally, toward knee brace

Knee-brace members are W8 × 20. From member property tables,

d = 8.14 in (206.7 mm) A = 5.89 in² (3800 mm²)
b = 5.268 in (133.8 mm) r_{min} = 1.25 in (31.75 mm)
t_w = 0.248 in (6.29 mm) I = 69.4 in⁴ (2.89 × 10⁷ mm⁴)
S = 17 in³ (2.79 × 10⁵ mm³)

The baseplates are 1 in × 14 in × 14 in (25.4 mm × 356 mm × 356 mm) with four 1-in (25.4-mm) diameter concrete anchor bolts with 6-in (152.4-mm) embedment. The bolt hole centers are 1.5 in (38.1 mm) from the edge of the plates. The restraint structure (knee brace) requires a minimum stiffness of 600,000 lb/in (105,120 N/mm). The restraint loads are ±4500 lb (20,025 N) OBE and ±9000 lb (40,050 N) SSE.

First the pipe local stress must be checked. Since there are no thermal or weight loads on a snubber, only the upset and emergency cases need to be checked. Since the attachment is a trunnion, WRC Bulletin 107 must be used. In this case, only a radial load is applied. Referring to Sec. 8.3.1.4, we find

$$S_{long} = \frac{C_1 P}{Rt} + \frac{C_2 6P}{t^2}$$

$$S_{circ} = \frac{C_7 P}{Rt} + \frac{C_8 6P}{t^2}$$

where R = (12.75 − 0.375)/2 = 6.1875 in (157.2 mm)
t = 0.375 in (9.525 mm)
r_o = 6.625/2 = 3.3125 in (84.1 mm)
β = 0.875(3.3125)/6.1875 = 0.875(84.1)/157.2 = 0.47
γ = 6.1875/0.375 = 157.2/9.525 = 16.5

Based on the above parameters, the constants can be found by consulting the figures published in WRC Bulletin 107:

C_1 = 1.8 C_7 = 1.0
C_2 = 0.016 C_8 = 0.03

and

S_{long} = 1.458P psi
       = (2.26 × 10⁻³)P N/mm²
S_{circ} = 1.711P psi
       = (2.65 × 10⁻³)P N/mm²

The allowable pipe stress for the upset condition (service level B) is 1.2S_h; for emergency (service level C) it is 1.8S_h. The stresses in the pipe wall are summarized in Table 8.5. Therefore the local stresses are acceptable.

The weld between the trunnion and the pipe must be checked:

A_w = $\pi(6.625)$ = 20.8 in (529 mm)
F_w = 9000/20.8 = 433 lb/in
    = 40,050/529 = 76 N/mm
F_w = 0.707 × 0.1875 × 0.3 × 70,000 = 2783 lb/in ≥ 433 lb/in
    = 0.707 × 4.76 × 0.3 × 483 = 488 N/mm ≥ 76 N/mm

TABLE 8.5 Stresses in Pipe Wall

Loading case	Longitudinal pipe stress	$+ S_{\text{long}}$	= Total	Circumferential pressure stress	$+ S_{\text{circ}}$	= Total	Allowable
Upset	8240 psi 57 N/mm²	6561 psi 45.3 N/mm²	14,801 psi 102.3 N/mm²	4800 psi 33.2 N/mm²	7700 psi 53.1 N/mm²	12,500 psi 86.3 N/mm²	18,000 psi 123.6 N/mm²
Emergency	12,580 psi 87 N/mm²	13,122 psi 90.6 N/mm²	25,702 psi 177.6 N/mm²	4800 psi 33.2 N/mm²	15,400 psi 106.2 N/mm²	20,200 psi 139.4 N/mm²	27,000 psi 185.4 N/mm²

Therefore the weld is adequate. The weld between the trunnion and the cover plate is acceptable by comparison. The trunnion must be checked. For 36-kips/in² (248-N/mm²) steel,

$$\frac{kl}{r} = 2.1(12)/2.245 = 11$$

$$= 2.1(305)/57 = 11$$

For $kl/r = 11$, $F_A = 21,100$ psi (145.5 N/mm²) for the upset case (per ASME code, Section III, Subsection NF). Since service level C stress allowances may be increased by 33 percent, in this case, the emergency load (twice the upset) will always govern. Therefore the structure need only be checked for the emergency load:

$$F_A = \frac{9000}{5.58} = 1613 \text{ psi} \leq 21,100 \text{ psi}$$

$$= \frac{40,050}{3600} = 11.1 \text{ N/mm}^2 \leq 145.5 \text{ N/mm}^2$$

The trunnion is acceptable. The snubber must be selected next, based on the LCDS shown in Fig. 8.24. Based on the maximum load of 9000 lb (40,050 N) and the thermal movement of 2 in (50.8 mm), a size 3 [rated for 10,000 lb (44,500 N)] with a total travel of 4 in (102 mm) may be used. The cold piston setting can be found by

$$\text{Cold setting} = \frac{TT + K\Delta_{\text{TH}}}{2}$$

where TT = total travel of snubber, in (mm)

K = +1 when snubber will retract, cold to hot; -1 when snubber will extend cold to hot

Δ_{TH} = expected thermal movement, in (mm)

In this the cold setting of the snubber comes to 3 in (76.2 mm). When it is retracted in the hot condition, the piston setting will be 1 in (25.4 mm). The weld for the snubber bracket is required by the vendor to be a ¼-in (6.35-mm) all-around fillet weld.

Next the knee brace must be checked. Figure 8.23b shows the results of a moment distribution analysis on the frame. The vertical member sees a maximum moment of 108,000 in·lb (1.22×10^7 mm·N) above the brace and a moment of 63,720 in·lb (7.2×10^6 mm·N) and an axial force of 14,786 lb (65,798 N) below the brace. The brace has a maximum moment of 44,280 in·lb (5×10^6 mm·N) and an axial force of $(12,414 + 14,786)/\sqrt{2} = 19,233$ lb (85,588 N). The maxi-

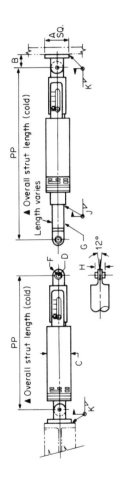

Size	Maximum load	Standard total travel	Minimum PP	Maximum PP	A	B	C (max)	D	F	G	H	J	K	Approximate weight
1	750	2	1'5"	6'0"	3	1½	2⅞	0.499	⅞	1¼	0.796	⅛	⅛	7
	3,338	50.8	432	1,830	76.2	38.1	73.0	12.7	22.2	Sch 40	19.9	3.2	3.2	31
		4	1'9"											
		101.6	533											
2	3,000	4	1'11"	6'0"	3	1½	3¾	0.499	⅞	1½	0.795	⅛	⅛	20
	13,350	101.6	584	1,830	76.2	38.1	95.3	12.7	22.2	Sch 40	19.9	3.2	3.2	89
		8	2'7"											
		203.2	787											
3	10,000	4	2'1"	8'0"	4	2½	4½	0.749	1¼	2	1.151	3/16	¼	35
	44,500	101.6	635	2,440	101.6	63.5	114.3	19.0	31.8	Sch 80	29.2	4.8	6.4	156
		8	2'9"											
		203.2	838											
4	20,000	6	2'10"	8'0"	5	3	5⅝	0.999	1½	2	1.536	3/16	¼	75
	89,000	152.4	864	2,440	127	76.2	142.9	25.4	38.1	Sch 80	39.0	4.8	6.4	334
		12	3'10"											
		304.8	1,168											

USCS: lb, ft, in.   SI: N, mm

Figure 8.24 Mechanical snubbers, load capacity ratings, and dimensions.

mum shear in any member is 9000 lb (40,050 N). By checking the W8 × 20, we find for a compact section of 36-kips/in² (248-N/mm²) steel

$$\tau_{max} = \frac{9000}{8.14 \times 0.258} = 4285 \text{ psi} \leq 1.33(0.4 \times 36{,}000) = 19{,}200 \text{ psi}$$

$$= \frac{40{,}050}{207 \times 6.3} = 30.7 \text{ N/mm}^2 \leq 1.33(0.4 \times 248) = 132.3 \text{ N/mm}^2$$

Above the brace,

$$\sigma_B = \frac{108{,}000}{17} = 6353 \text{ psi} \leq 1.33 \times 0.66 \times 36{,}000 = 31{,}680 \text{ psi}$$

$$= \frac{1.22 \times 10^7}{2.79 \times 10^5} = 43.7 \text{ N/mm}^2 \leq 1.33 \times 0.66 \times 248 = 218 \text{ N/mm}^2$$

Below the brace,

$$\frac{Kl}{r} = \frac{1.2(28)}{1.25} = 27$$

$$= \frac{1.2(711)}{31.75} = 27$$

$$F_a = 20{,}150 \times 1.33 = 26{,}867 \text{ psi } (185.3 \text{ N/mm}^2)$$

Since $f_a/F_a = (14{,}786/5.89)/26{,}867 = 0.09 \leq 0.15$,

$$\frac{14{,}786/5.89}{26{,}867} + \frac{63{,}720/17}{31{,}680} = 0.21 \leq 1.0$$

or

$$\frac{65{,}798/3800}{185.3} + \frac{7.2 \times 10^6/(2.79 \times 10^5)}{218} = 0.21 \leq 1.0$$

Since the earthquake load is reversible, the brace will also see compression:

$$\frac{Kl}{r} = \frac{1.2(28\sqrt{2})}{1.25} = 38$$

$$= \frac{1.2(711\sqrt{2})}{31.75} = 38$$

$$F_a = 19{,}350 \times 1.33 = 25{,}800 \text{ psi } (178 \text{ N/mm}^2)$$

Since $f_a/F_a = (19{,}233/5.89)/25{,}800 = 0.13 \leq 0.15$, then

$$\frac{19{,}233/5.89}{25{,}800} + \frac{44{,}280/17}{31{,}680} = 0.21 \leq 1.0$$

$$\frac{85{,}588/3800}{178} + \frac{5 \times 10^6/(2.79 \times 10^5)}{218} = 0.21 \leq 1.0$$

Therefore the steel in the knee brace is adequate. The welds must be checked. Since all welds in the structure are of the same pattern (conservatively neglecting elongations due to angular incidences), they may all be qualified by checking one for a conservative load combination. For example, the greatest loads on any of the welds are bending of 44,280 in·lb (5 × 10⁶ mm·N), tension of 14,786 lb (65,798 N), and shear force of 14,786 lb (65,798 N). The weld group properties are, conservatively,

$$A_w = 4(5.268) + 2(8.14) = 37.35 \text{ in } (949 \text{ mm})$$

$$S_w = \frac{2(5.268 \times 8.14) + (8.14)^2}{3} = 107.8 \text{ in}^2 \ (6.96 \times 10^4 \text{ mm}^2)$$

$$f_w \leq \sqrt{\left(\frac{14{,}786}{37.35} + \frac{44{,}280}{107.8}\right)^2 + \left(\frac{14{,}786}{37.35}\right)^2} = 899 \text{ lb/in}$$

$$\leq \sqrt{\left(\frac{65{,}798}{949} + \frac{5 \times 10^6}{6.96 \times 10^4}\right)^2 + \left(\frac{65{,}798}{949}\right)^2} = 157 \text{ N/mm}$$

$$F_{w,\ min} = 0.707 \times 0.25 \times 0.3 \times 70{,}000 = 3712 \text{ lb/in} \geq 899 \text{ lb/in}$$

$$= 0.707 \times 6.4 \times 0.3 \times 483 = 656 \text{ N/mm} \geq 157 \text{ N/mm}$$

Therefore, all the welds in the frame are adequate by comparison to this calculation.

The baseplates must now be evaluated. They are loaded with moment, tension, and shear. Again, as with the welds, a single calculation with a conservative load combination may be done to qualify both baseplates. The maximum loads on the plates are a moment of 31,860 in · lb (3.6 × 10^6 mm · N), a tension of 14,786 lb (65,798 N), and a shear force of 12,414 lb (55,242 N). From Sec. 8.3.4.2, assuming baseplate rigidity, the pivot point may be located:

$$Y = \frac{\sqrt{A_B^2(E_s/E_c)^2 + 2A_B(E_s/E_c)D\omega} - A_B(E_s/E_c)}{W} \quad \text{in (mm)}$$

For this problem, use $E_s = 28 \times 10^6$ psi (1.93 × 10^5 N/mm^2) and $E_c = 3.6 \times 10^6$ psi (2.51 × 10^4 N/mm^2). For two 1-in (25.4-mm) bolts in tension, $A_B = 2\pi (0.5)^2 = 1.57$ in^2 (1013 mm^2), and $D = 14 - 1.5 = 12.5$ in (317.5 mm). Therefore

$$Y = \frac{\sqrt{(1.57)^2(28/3.6)^2 - 2(1.57)(28/3.6)(12.5)(14)} - (1.57)(28/3.6)}{14} = 3.88 \text{ in}$$

$$= \frac{\sqrt{(1013)^2(19.3/2.51)^2 + 2(1013)(19.3/2.51)(317.5)(356)} - (1013)(19.3/2.51)}{356} = 98 \text{ mm}$$

The maximum tension per bolt (without prying) due to the applied moment is

$$F_T = \frac{31{,}860}{2(12.5 - 3.88/3)} = 1421 \text{ lb per bolt}$$

$$= \frac{3.6 \times 10^6}{2(317.5 - 98/3)} = 6319 \text{ N per bolt}$$

The applied tension load may be distributed between the bolts [1.5 in (38.1 mm) from the plate edge] and the concrete compression [1.29 in (32.7 mm) from the plate edge]:

$$F_T = \frac{14786(7 - 1.29)}{2(14 - 1.5 - 1.29)} = 3766 \text{ lb per bolt}$$

$$= \frac{65{,}798(178 - 32.7)}{2(356 - 38.1 - 32.7)} = 16{,}761 \text{ N per bolt}$$

Considering prying, we see the total bolt tension is

$$F_p = (1421 + 3766)\left[1 + \frac{100(1.5)(1)^2 - (\tfrac{1}{2})(18)(14)(1)^2}{70(1.5)(1)^2 + (\tfrac{1}{2})(21)(14)(1)^2}\right] = 5681 \text{ lb/bolt}$$

$$= (6319 + 16{,}761)\left[1 + \frac{100(38.1)(25.4)^2 - \tfrac{1}{2}(18)(356)(25.4)^2}{70(38.1)(25.4)^2 + \tfrac{1}{2}(21)(356)(25.4)^2}\right] = 25{,}264 \text{ N/bolt}$$

Shear is distributed equally among the four bolts:

$$F_T = \frac{12{,}414}{4} = 3104 \text{ lb/bolt } (13{,}811 \text{ N/bolt})$$

The anchor bolts themselves may now be checked. The tension reduction factor (RF) based on a single bolt interference (the other row of bolts is in the compression area) at a distance of 11 in (279.4 mm) and an embedment of 6 in (152.4 mm), is calculated as:

$$\text{RF} = 1 - \left[\frac{2\cos^{-1}(5.5/6)}{360} - \frac{11/2\sqrt{6^2 - (11/2)^2}}{6^2}\right] = 0.986$$

$$= 1 - \left\{\frac{2\cos^{-1}[(279.4/2)/152.4]}{360} - \frac{(279.9/2)\sqrt{(152.4)^2 - (279.4/2)^2}}{\pi(152.4)^2}\right\} = 0.986$$

By using the ultimate tension and shear values for 1-in (25.4-mm) diameter anchor bolts in 4000-psi (27.6-N/mm²) concrete shown in Table 8.6, a squared interaction formula, and a factor of safety of 4, the anchor bolts may be checked. From Table 8.6, $T_u = 24{,}941$ lb (110,987 N) and $v_u = 34{,}491$ lb (153,485 N).

$$\left(\frac{4 \times 5681}{0.986 \times 24{,}941}\right)^2 + \left(\frac{4 \times 3104}{34{,}491}\right)^2 = 0.98 \leq 1.0$$

$$\left(\frac{4 \times 25{,}264}{0.986 \times 110{,}987}\right)^2 + \left(\frac{4 \times 13{,}811}{153{,}485}\right)^2 = 0.98 \leq 1.0$$

Plate bending stresses must be checked now. The allowable stress for plates bent about their weak axis is $0.75S_y$. This may be increased by 1.33 for emergency

TABLE 8.6 Average Ultimate Tensile and Shear Loads for 1-in (25.4-mm) Diameter Wedge Anchor

Anchor embedment depth, in (cm)	Concrete Strength					
	2,000 psi (14,000 kPa)		4,000 psi (28,000 kPa)		6,000 psi (42,000 kPa)	
	Ultimate tension	Ultimate shear	Ultimate tension	Ultimate shear	Ultimate tension	Ultimate shear
4.5 (11.4)	14,000 (62,300)	27,355 (121,730)	16,000 (71,200)	26,879 (119,612)	20,500 (91,225)	32,112 (142,898)
5 (12.7)	15,500 (68,975)	27,355 (121,730)	18,900 (84,105)	26,879 (119,612)	24,941 (110,987)	32,112 (142,898)
6 (15.2)	17,600 (78,320)	27,355 (121,730)	24,941 (110,987)	34,491 (153,485)	24,941 (110,987)	36,394 (161,953)
7 (17.8)	18,200 (80,990)	27,355 (121,730)	24,941 (110,987)	34,491 (153,485)	24,941 (110,987)	36,394 (161,953)
8 (20.3)	18,200 (80,990)	27,355 (121,730)	24,941 (110,987)	34,491 (153,485)	24,941 (110,987)	36,394 (161,953)
9 (22.9)	18,200 (80,990)	27,355 (121,730)	24,941 (110,987)	34,491 (153,485)	24,941 (110,987)	36,394 (161,953)
10 (25.4)	18,200 (80,990)	27,355 (121,730)	24,941 (110,987)	34,491 (153,485)	24,491 (110,987)	36,394 (161,953)

loading. Plate bending stress is estimated as

$$\sigma_{Pl} = \frac{6nF_pb}{wt^2} = \frac{6(2)(5681)(1.5)}{(14)(1)^2} = 7304 \text{ psi} \leq 36{,}000 \text{ psi}$$

$$= \frac{6(2)(25{,}264)(38.1)}{356(25.4)^2} = 50.3 \text{ N/mm}^2 \leq 248 \text{ N/mm}^2$$

Both baseplates may be qualified by comparison to this calculation with conservative loads. This calculation completes the evaluation of the support from a load capacity point of view. However, the stiffness of the restraint must still be checked against the required stiffness. This is done by first finding the displacement under load. The displacement can be found by taking the double integral of the moment divided by EI over the length of the vertical member. The moment diagram for this member is shown in Fig. 8.25. The rotation at the point of load can be found by

$$\theta = \frac{-31{,}860(40)}{EI} + \frac{3413(40)^2}{2EI} + \frac{44{,}280(12)}{EI} - \frac{12{,}414(12)^2}{2EI}$$

$$= -\frac{(3.6 \times 10^6)(1016)}{EI} + \frac{15{,}188(1016)^2}{2EI} + \frac{(5 \times 10^6)(305)}{EI} - \frac{55{,}242(305)^2}{2EI}$$

Integrating further, the displacement is calculated as

$$\Delta = \frac{-31{,}860(40)^2}{2EI} + \frac{3413(40)^3}{6EI} + \frac{44{,}280(12)^2}{2EI} - \frac{12{,}414(12)^3}{6EI} = \frac{1.05 \times 10^7}{EI}$$

$$= \frac{-(3.6 \times 10^6)(1016)^2}{2EI} + \frac{15{,}188(1016)^3}{6EI} + \frac{(5 \times 10^6)(305)^2}{2EI} - \frac{55{,}242(350)^3}{6EI}$$

$$= \frac{7.68 \times 10^{11}}{EI}$$

For $E = 28 \times 10^6$ psi (1.93×10^5 N/mm^2) and $I = 69.4$ in^4 (2.89×10^7 mm^4), the displacement is

$$\Delta = \frac{1.05 \times 10^7}{(28 \times 10^6)(69.4)} = 5.4 \times 10^{-3} \text{ in}$$

$$= \frac{7.68 \times 10^{11}}{(1.93 \times 10^5)(2.89 \times 10^7)} = 0.138 \text{ mm}$$

The restraint stiffness (spring constant) is found by dividing the applied load by the dispacement under that load:

$$K = \frac{9000}{5.4 \times 10^{-3}} = 1.67 \times 10^6 \text{ lb/in} \geq 600{,}000 \text{ lb/in} \quad \text{(required minimum stiffness)}$$

$$= \frac{40{,}050}{0.138} = 290{,}217 \text{ N/mm} \geq 105{,}120 \text{ N/mm}$$

Therefore the restraint is also suitable from a stiffness point of view.

Problem 8.6 Figure 8.26 shows a pipe passing through a wall penetration in a building. Lugs and shim plates have been used to turn the penetration into a lateral penetration and a vertical support. The following information is given:

Piping code: B31.1
Pipe size: 6-in (150-mm) nominal diameter, schedule 80 pipe; $D = 6.625$ in (168.3 mm), $t = 0.432$ in (11 mm)

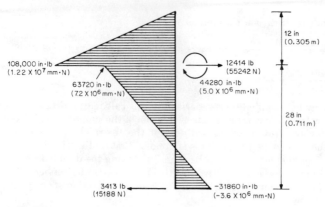

Figure 8.25 Moment diagram for vertical member.

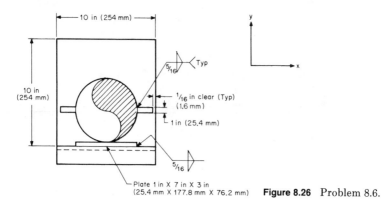

Figure 8.26 Problem 8.6.

Pipe material allowable stress:	A106 GrB carbon steel, $S_A = 22{,}500$ psi (155 N/mm^2), $S_h = 15{,}000$ psi (103 N/mm^2)
Existing pipe stresses:	thermal: 3400 psi (23.5 N/mm^2); sustained: 4680 psi (32.3 N/mm^2); circumferential pressure stress = 4200 psi (29 N/mm^2); longitudinal pressure stress = 2100 psi (14.5 N/mm^2)
Lug size:	1 in (25.4 mm) × 1.625 in (41.3 mm) × 2 in (50.8 mm)
Line contact length:	3 in (76.2 mm)
Design loads:	Lateral (x): thermal = 3050 lb (13,573 N), deadweight = 20 lb (89 N); vertical (y): thermal = 1100 lb (4895 N), deadweight = -1700 lb (-7565 N)

Consider friction, using a friction coefficient of 0.4.

The first step is to determine the design load combination. The pipe can be in the operating (hot) condition or in the nonoperating (cold) condition. The possible loads on the pipe during these two conditions are as follows:

Manual Calculation Methods 269

Operating condition	F_x	F_y
Cold (deadweight only)	20 lb (89 N)	−1700 lb (−7565 N)
Hot (thermal plus deadweight)	3070 lb (13,662 N)	−600 lb (−2670 N)

Friction should be considered whenever a pipe rubs directly on the support structure and there are significant thermal pipe movements.

The friction force is applied in the direction of the movement and is calculated as the maximum static (deadweight or deadweight plus thermal) force on the support times the friction coefficient. In this case, the friction force is in the z-direction and has a magnitude of $0.4(3070) = 1228$ lb (5465 N) at the lug-penetration interface and $0.4(1700) = 680$ lb (3026 N) at the interface of the pipe and shim plate. When local stress calculations are done, the friction force should be included with the thermal load case.

First the local stress must checked. The x-direction force is resisted by a lug, and the y-direction force is resisted by line contact. The lug will impart a radial load and a longitudinal moment (from the frictional force). From Sec. 8.3.1.3,

$$S_P = C_P\left(\frac{P}{A}\right) \qquad S_L = C_L\left(\frac{M_L}{Z}\right)$$

For the 6-in (150-mm) schedule 80 pipe, $A = 8.4$ in^2 (5419 mm^2), $Z = 12.23$ in^3 (2×10^5 mm^3), $R = 3.0965$ in (78.6 mm), and $t = 0.432$ in (11 mm).

$$C_P = 12\left(\frac{R}{t}\right)^{0.64} \eta^{1.54}$$

where $\eta = -(x_1 \cos\theta + y_1 \sin\theta) - \dfrac{1}{A_0}(x_1 \sin\theta - y_1 \cos\theta)^2$

$x_1 = x_0 + \log\beta_1$

$y_1 = y_0 + \log\beta_2$

$\beta_1 = L_1/2R = \dfrac{1}{2 \times 3.0965} = 0.161$

$\beta_2 = L_2/2R = \dfrac{2}{2 \times 3.0965} = 0.323$

$x_0 = 0$

$y_0 = 0.05$

$\theta = 40°$

$A_0 = 2.2$

Therefore, $\eta = 0.834$, and $C_P = 32$. So $S_P = 32(P/8.4) = 3.81P$ psi, or $32(P/5419) = 0.00591P$ N/mm^2. For the longitudinal moment,

$$S_L = C_L\left(\frac{M_L}{Z}\right) = C_L\left(\frac{F_z \times 1.625}{Z}\right)$$

$$C_L = 1.2\left(\frac{R}{t}\right)^{0.74}\eta^{4.74}$$

$\eta = -(x_1 \cos\theta + y_1 \sin\theta) - \dfrac{1}{A_0}(x_1 \sin\theta - y_1 \cos\theta)^2$

$x_0 = -0.45$

$y_0 = -0.55$

$\theta = 50°$

$A_0 = 2.0$

Therefore, $\eta = 1.556$, and $C_L = 41.9$. So $S_L = 41.9(F_z \times 1.625/12.23) = 5.57F_z$ psi or $41.9[F_Z \times 41.3/(2 \times 10^5)] = 0.0085F_z = $ N/mm². Again, the stresses induced by the friction force must be considered with the thermal loading case.

The stresses due to the line contact (see Sec. 8.3.1.2) must be computed in both the longitudinal and circumferential directions. For the longitudinal stress,

$$S_{\text{long}} = \left(\frac{0.52R^{1/4}}{L^{1/2}t^{7/4}}\right)P = \left[\frac{0.52(3.0965)^{1/4}}{3^{1/2}(0.432)^{7/4}}\right]F_y = 1.73F_y \quad \text{psi}$$

$$= \frac{52(78.7)^{1/4}F_y}{(76.2)^{1/2}(11)^{7/4}} = 0.00267F_y \quad \text{N/mm}^2$$

For the circumferential stress,

$$S_{\text{circ}} = \left(\frac{1.05R^{1/4}}{L^{1/2}t^{7/4}} + \frac{0.496R^{3/4}}{L^{3/2}t^{5/4}}\right)P = \left[\frac{1.05(3.0965)^{1/4}}{3^{1/2}(0.432)^{7/4}} + \frac{0.496(3.0965)^{3/4}}{3^{3/2}(0.432)^{5/4}}\right]F_y = 4.13F_y \quad \text{psi}$$

$$= \left[\frac{1.05(78.7)^{1/4}}{(76.2)^{1/2}11^{7/4}} + \frac{0.496(78.7)^{3/4}}{(76.2)^{3/2}11^{5/4}}\right]F_y = 0.00638F_y \quad (\text{N/mm}^2)$$

A stress summary is shown in Table 8.7. When the local pipe stresses are summed, the pressure stresses should be included only with primary stress cases, not with expansion stresses such as during the thermal load case. Therefore the local stresses are acceptable. Next the weld between the lug and the pipe must be checked. The weld properties are

$$A_w = 4 \text{ in } (101.6 \text{ mm}) \qquad S_w = 1.33 \text{ in}^2 \ (858.1 \text{ mm}^2)$$

Using the maximum loading on the lug (the deadweight plus thermal loading case) yields

$$F_w = \left[\left(\frac{3070}{4} + \frac{1228 \times 1.625}{1.33}\right)^2 + \left(\frac{1228}{4}\right)^2\right]^{1/2} = 2289 \text{ lb/in}$$

$$= \left[\left(\frac{13,362}{101.6} + \frac{5465 \times 41.3}{858.1}\right)^2 + \left(\frac{5465}{101.6}\right)^2\right]^{1/2} = 401 \text{ N/mm}$$

TABLE 8.7 Stress Summary

Load case	Moment stress	Longitudinal pressure stress	Lug stress F_x	Lug stress F_z	Total	Longitudinal line stress
TH	3400	—	11,621	6840	18,461†	1903
	23.5	—	80.2	47.3	127.5†	13.1
DW + PR	4680	2100	76	0	76	2941
	32.3	14.5	0.5	0	0.5	20.2

Load case	Circumferential pressure stress	Circumferential line stress	Maximum local stress*	Allowable
TH	—	4543	21,861	22,500 psi
	—	31.2	151	155 N/mm²
DW + PR	4200	7021†	11,221	15,000 psi
	29	48.3†	77.3	103 N/mm²

† Governing local stress.
* Maximum local stress is greatest of: (1) moment stress + longitudinal pressure stress + lug stress; (2) moment stress + longitudinal pressure stress + longitudinal line stress; (3) circumferential pressure stress + circumferential line stress.

$$w_{req} = \frac{2289}{0.707 \times 21,000} = 0.15 \text{ in}$$

$$= \frac{401}{0.707 \times 145} = 3.9 \text{ mm}$$

The $\tfrac{5}{16}$-in (7.9-mm) fillet weld is adequate. Next the lugs themselves should be checked. The properties of the lugs are

$$I_{min} = \frac{2(1)^3}{12} = 0.1667 \text{ in}^4 \ (6.94 \times 10^4 \text{ mm}^4)$$

$$S = \frac{(1)(2)^2}{6} = 0.667 \text{ in}^3 \ (1.09 \times 10^4 \text{ mm}^3)$$

$$A = (1)(2) = 2 \text{ in}^2 \ (1290 \text{ mm}^2)$$

$$r_{min} = \sqrt{\frac{I_{min}}{A}} = \sqrt{\frac{0.1667}{2}} = 0.289 \text{ in } (7.3 \text{ mm})$$

$$\frac{kl}{r} = \frac{2.1(1.625)}{0.289} = 12$$

$$F_a = 21,050 \text{ psi } (145 \text{ N/mm}^2)$$

Since $(3070/2)/21,050 = 0.07 \leq 0.15$,

$$\frac{3070/2}{21,050} + \frac{1228 \times 1.625/0.667}{0.6(36,000)} = 0.21 \leq 1.0$$

or

$$\frac{13,362/1290}{145} + \frac{5465 \times 41.3(1.09 \times 10^4)}{0.6(248.3)} = 0.21 \leq 1.0$$

$$\tau = \frac{1228}{2} = 614 \text{ psi} \leq 0.4(36,000) = 14,400 \text{ psi}$$

$$= \frac{5465}{1290} = 4.2 \text{ N/mm}^2 \leq 0.4(248.3) = 99.3 \text{ N/mm}^2$$

The friction force at the point where the pipe rubs against the shim plate has no noticeable effect on the pipe, since it causes no bending stresses on the pipe wall. The shim plate sees no bending due to the y-direction force since the load is transferred directly to the wall through the bearing. The frictional force will cause both bending and torsion on the plate and weld, however:

$$M_{y,max} = \frac{680(7)}{8} = 595 \text{ in} \cdot \text{lb}$$

$$= \frac{3026(177.8)}{8} = 67,253 \text{ mm} \cdot \text{N}$$

$$M_{x,max} = \frac{680(0.5)}{2} = 170 \text{ in} \cdot \text{lb}$$

$$= \frac{3026(12.7)}{2} = 19,215 \text{ mm} \cdot \text{N}$$

$$V_{max} = \frac{680}{2} = 340 \text{ lb}$$

$$= \frac{3026}{2} = 1513 \text{ N}$$

Checking the plate in bending and shear reveals

$$S = \frac{(1)(3)^2}{6} = 1.5 \text{ in}^3$$

$$= \frac{(25.4)(76.2)^2}{6} = 2.46 \times 10^4 \text{ mm}^3$$

$$\sigma = \frac{595}{1.5} = 397 \text{ psi} \leq 0.6(36{,}000) \text{ psi}$$

$$= \frac{67{,}253}{2.46 \times 10^4} = 2.7 \text{ N/mm}^2 \leq 0.6(248.3) \text{ N/mm}^2$$

$$R = \frac{(3)(1)^3}{3} = 1 \text{ in}^4$$

$$= \frac{(75.2)(25.4)^3}{3} = 4.1 \times 10^5 \text{ mm}^4$$

$$A = 1(3) = 3 \text{ in}^2$$

$$= 25.4(76.2) = 1935.5 \text{ mm}^2$$

$$\tau = \frac{340}{3} + \frac{170(1)}{1} = 283 \text{ psi} \leq 0.4(36{,}000) \text{ psi}$$

$$= \frac{1513}{1935.5} + \frac{19{,}215(25.4)}{4.1 \times 10^5} = 2 \text{ N/mm}^2 \leq 0.4(248.3) \text{ N/mm}^2$$

Therefore the plate is adequate. Checking the weld at each end of the plate shows

$$A_w = 3 \text{ in } (76.2 \text{ mm})$$

$$S_w = \frac{(3)^2}{6} = 1.5 \text{ in}^2 \ (968 \text{ mm}^2)$$

$$F_w = \left[\left(\frac{595}{1.5}\right)^2 + \left(\frac{680 \times 1}{2(1.5)}\right)^2 + \left(\frac{340}{3}\right)^2\right]^{1/2} = 471 \text{ lb/in}$$

$$= \left[\left(\frac{67{,}253}{968}\right)^2 + \left(\frac{3026 \times 25.4}{2(968)}\right)^2 + \left(\frac{1513}{76.2}\right)^2\right]^{1/2} = 82 \text{ N/mm}$$

$$w_{req} = \frac{471}{0.707 \times 21{,}000} = 0.03 \text{ in}$$

$$= \frac{82}{0.707 \times 145} = 0.8 \text{ mm}$$

The existing $\frac{5}{16}$-in (7.9-mm) fillet weld is adequate, so the restraint is acceptable as designed.

Chapter 9

Computer Applications for Design and Analysis

9.1 Introduction

The computer, as a tool for design and engineering, has undergone an evolutionary process that spans over a quarter of a century. The computer was adapted for engineering use during the mid 1950s. The introduction of the computer initiated the period when semiautomated problem-solving processes began to supplant manual methods, especially for difficult and repetitive calculations. In the next decade the computer was perfected as a design tool for the aerospace industry. During this period, the computer demonstrated its ability to provide design services that were not cost-effective when performed by humans. For instance, computers were used to solve complex problems involving heat transfer, stress analysis, and vibration. In the 1970s and 1980s, the computer revolution continued with the introduction of new applications and the increased accessibility of mini- and microcomputers. The computer provides services with greater speed, accuracy, and less cost than was even considered imaginable just a few decades ago.

An encouraging aspect of the computer industry is that the cost of computer equipment continues to drop as the computer's ability continues to increase. This phenomenon is a result of the great technological progress made since the early days of the computer—many new companies can now provide hardware and software based on the research and development of the industry pioneers.

In this chapter we discuss first the computer itself and then such applications as computer-aided design and drafting (CADD) as well as computer-aided engineering (CAE). A brief history of the development of these systems is presented, with the emphasis on those aspects pertaining to pipe support design, such as stress analysis and structural design programs.

It is important to remember that by automating the design, drafting, and engineering process, the computer does not dehumanize the task being accomplished. Instead, the computer provides a tool to do repetitive tasks in a faster, less expensive, and more accurate way and thus extends the user's ability. The added speed of execution also provides answers in seconds that previously could have taken months to develop. Therefore the computer extends our ability to solve problems that were not previously feasible.

Further, note that the computer is not always "right." Despite the high level of accuracy of computerized calculations, the result is only as good as the input data. The engineer's skills are still needed in order to provide the computer program with an accurate model of the problem to be solved and to interpret the results.

9.1.1 Introduction to the computer

The computer is becoming the central tool of today's engineer. Most engineers spend a great deal of time processing information. The engineer must read, write, calculate, design, plan, and communicate. All these tasks can be aided by the computer.

There are four major categories of digital computers: super computers, mainframe computers, minicomputers, and microcomputers. A personal computer is usually classified as a general-purpose microcomputer. The main differences between these categories are in data capacity, computing speed, and utility. Table 9.1 presents a comparative chart of typical computer characteristics. Figure 9.1 shows typical examples of these types of computers.

The growth in the popularity of the microcomputer, first introduced in 1975, during the 1980s shows the result of what has been called the "silicon revolution." During this period the memory size per chip increased by at least a factor of 50, reducing the computer hardware to desktop size and bringing the cost within the reach of most businesses and many consumers. These computers are transforming the way engineering groups operate in their daily work habits. The number of applications seems to be limited only by human ingenuity.

The computer is a system comprised of a logical interconnection of fundamental units, each having a specific purpose. The units are called *subsystems,* and these are composed of even smaller units. Each unit has a specific task or purpose. There are two major types of computers: analog and digital. These computers vary in their construction, method of data manipulation, and the means by which they respond to instructions. An analog computer works by mimicking whatever it is computing; it does

TABLE 9.1 Comparison of Computer Characteristics

Characteristic	Computer		
	Mainframe	Mini	Micro
Size:			
Physical size	Room-sized	Desk-sized	Typewriter-sized
Word size	32 binary digits	16 binary digits	8 or 16 binary digits
Memory size	10,000 Kilobytes	500 Kilobytes	250 Kilobytes
CPU cycle time	80 ns	200 ns	1 μs
Memory cycle time	500 ns	800 ns	1 μs
Cost:	$2 million to $4 million	$20,000 to $40,000	$2000 to $4000
Utility:			
Operating system	Multiprogramming	Multiprogramming	Single task
Languages supported	Most common	Some	A few
Operating requirements	Special environment	Limited need for operator	User-operated
Applications packages	Excellent	Good	Limited at present

this by varying its own physical properties in a continuous way. In analog computers, circuits are programmed by presetting the physical characteristics of the computer components, such as variable resistors, capacitors, and inductors. The analog computer has a primary component called an *operational amplifier.* These amplifiers perform mathematical functions such as summation, differentiation, and integration. These calculations are performed by summing the input voltage, determining the rate of change (differential) of the input voltage, or by integrating over time one or more input voltages. The operational amplifiers can also invert the input voltage to produce a negative output. Analog computers are usually used for scientific research purposes and some mechanical applications, such as in the automatic transmissions of automobiles. There are few large general-purpose analog computers in use.

Digital computers operate by performing arithmetic functions on data which is in numerical form. Modern digital computers go through five steps during the solution of problems. The first step involves the input of data which includes instructions and associated information. The second step is the memory step that considers where the data and instructions should be stored. The third step, a control step, considers the activities (and their relation to each other) required to solve the problems according to the prescribed instructions. The fourth step is the execution of the instructions, in the form of logical operations, calculations, or data manipulation. Finally, the fifth step is the output step, where the computational results and other data are displayed to the user. If the "user" is another

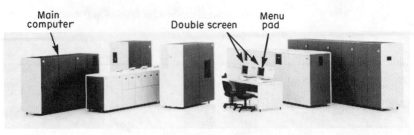

(a)

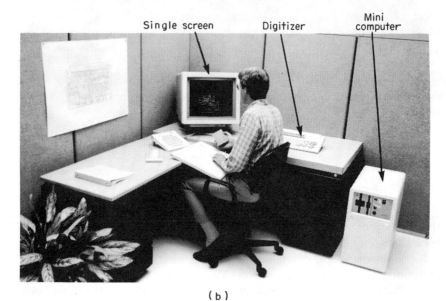

(b)

(c)

Figure 9.1 Examples of (a) mainframe; (b) minicomputer; (c) microcomputer. *(Courtesy of International Business Machines Corporation.)*

machine, the output could be in the form of data on tapes, disks, punched cards, or electronic signals, while humans require something that is in an easily understood form.

The most important characteristics of digital computers are speed, ability to operate in exactly repeatable ways, and flexibility. Digital computers operate in the binary system, using on-off switches to represent the binary digits, or bits. The binary numbers in the digital computer can be used to represent letters and symbols, as well as numbers, when the computer is in the character mode.

9.1.2 Input/output devices

A computer must communicate with its users in order to receive information. Computers may also need to communicate with operators, actuators, or other computers. The input data must be translated to a form which can be read by the computer. The most common types of input media are punched cards or paper tape, magnetic tape, magnetic disks, drums, and manual interaction through a terminal keyboard. Input devices such as optical scanners, magnetic scanners, and voice input are less widespread, but are expected to be more prevalent in the future. These input devices are briefly discussed in the following sections. See Figs. 9.2 and 9.3 for the identification of some of the components.

Punched cards and paper tapes were the most widely used methods of data entry in the 1960s and 1970s. Punched cards can be used where the data can be divided into units that fit easily onto a single input line. Input is punched onto cards by a keyboard-type machine similar to a typewriter. When a character is typed, a pattern of holes is punched in a column of the card. A machine called a *card reader* interprets these cards one at a time and translates the patterns on the cards to electric pulses. The cards are arranged with 80 columns, thus permitting each card to carry data lines up to 80 characters long.

Paper tape is similar to punched cards in that both require rows of holes punched in paper. The tape is used where the data is of a continuous nature, such as with stored data that does not require additions or changes. Data is punched on tapes by punch tape devices which are connected either to a computer or to a keyboard.

Paper tape size is usually classified as either five-channel or eight-channel, both having their own character codes. Characters are represented by hole patterns, as with the punched cards. The reading devices for the tape and cards are similar.

Magnetic media are used to store large quantities of data for both input and output purposes as well as for what is usually called *secondary memory*. The secondary memory is a cheaper form of memory than the random-access memory (RAM). Intermediate data can be stored here if the

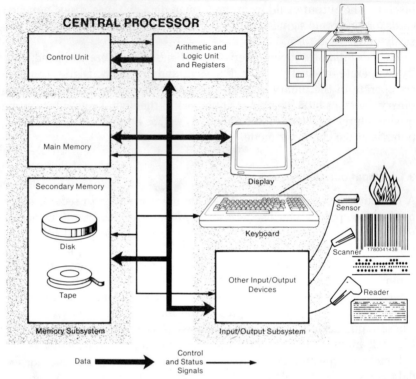

Figure 9.2 Major components of computer system. *(Courtesy of Engineering Planning and Management, Inc.)*

computer lacks sufficient RAM to hold all the data but needs to work on only portions of the data at any given time. A major consideration in using secondary memory is that the data must be transmitted to and from primary memory before and after each use, which adds time to the program execution. The cost of primary memory versus execution time is a trade-off that should be considered in any computer use efficiency model.

Magnetic media include magnetic tape, magnetic drums, and magnetic disks. All these work like sound tape recordings, except that the information conforms to electronic impulses rather than sound waves. To use this medium, the data must be converted to numerical data conforming to the binary digits of 0 and 1. Tapes are used more efficiently if the data is continuous since it is time-consuming to search through the tape to find specific pieces of data. When the data is continuous, the tapes provide a compact storage medium that can pack data quite densely, especially compared to the space required to store punched cards. Magnetic hard disks and drums offer higher-speed access than tapes and are usually used when it is necessary to search for data frequently.

A floppy disk is comprised of a thin, flexible piece of plastic or Mylar that is coated on one or both sides by magnetic material. The disk is encased in a protective envelope that protects against scratching or fingerprints and contains holes that allow access by the computer's reading devices, known as *disk drives*.

Floppy disks require a magnetic head read (as do tapes) rather than a needle, as used for audio disks. The floppy disk is erasable; therefore, it is cautioned that the contents might be altered or lost if the disk is not handled properly. In the United States, floppy disks are usually manufactured in three sizes: $3\frac{1}{2}$, $5\frac{1}{4}$, and 8 in (8.9, 13.3, and 20.3 cm) in diameter.

Hard (or fixed) disks are quite similar to floppy disks. The major difference is that the hard disks are made of nonflexible metal material. They are enclosed in airtight, dustproof housings. The hard disk can transmit data much more quickly than magnetic tape or floppy disks.

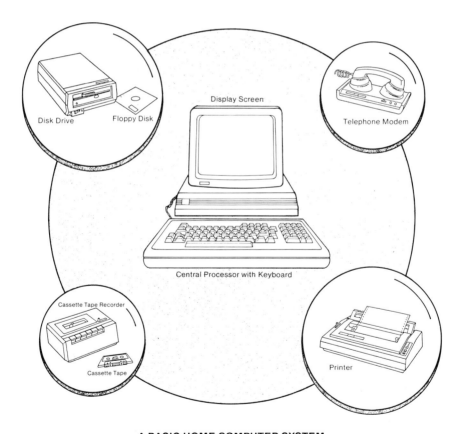

A BASIC HOME COMPUTER SYSTEM

Figure 9.3 Examples of microcomputer hardware. *(Courtesy of Engineering Planning and Management, Inc.)*

Time required for searching for data on a hard disk is usually measured in fractions of seconds, while the floppy disk may take several seconds and the magnetic tape could take minutes. Hard disks are usually used on larger computers and minicomputers while floppy disks are more often found in mocrocomputers, personal computers, and word processing machines.

Optical disks, like magnetic disks, store data on concentric rings. The optical disk requires the computer to have a device that can read little "pits" that correspond to binary digits. The variations in intensity of a laser light reflected from the pits are detected by a photocell that converts them into the type of electric signal required by the computer.

The keyboard of a computer terminal (see Figs. 9.2 and 9.3) looks like a typewriter keyboard, but there are more keys and so more input items. The input instructions and data can be entered by use of the keyboard with a cathode-ray tube (CRT) screen, which displays the data as it is typed. Using the screen and keyboard, the operator can interact with the computer program as it is being run or enter data to secondary storage devices, bypassing media such as punched cards or tape.

The data entered from the keyboard may be transmitted directly to the computer or to some intermediate storage device until all information has been entered. At this point, the entire data file may be transmitted as a whole to the computer.

Printers are used to provide a permanent copy of the computer output. Printers may range from electric typewriters that print one character at a time at rates of 15 characters per second (15 lines per minute) to mechanical printers capable of printing 132 characters on a line at a rate of up to 2000 lines per minute. Electrographic laser printers are capable of continuous paper speed of 31 in/s (787 mm/s) and up to 20,000 lines per minute.

Even the speeds of the fastest printers are slow compared to the computer's ability to generate this data. Therefore data must usually be stored in memory until the printer is ready.

Devices within the computer system may be interconnected through ordinary telephone lines by a device known as a *modem (modulator-demodulator)*. Modems can be of two different designs. One design, shown in Fig. 9.3, connects the telephone receiver temporarily to the computer through a boxlike connection. The second type of modem uses a permanent connection to the telephone line, thus maintaining continuous access. The first type of modem is more portable owing to the quick connect and disconnect features.

9.1.3 Central processing unit

Two essential subsystems of a computer are the control unit (CU) and the arithmetic and logic unit (ALU). Together these units comprise the central processing unit (CPU). The CPU performs all the arithmetic and logical operations, manipulation of data, and coordination of instructions required by the computer program.

In modern computer systems, some of the processing, storage, and input and output equipment can be located at different sites. Through data transmission equipment such as a telephone modem, terminals, satellite communications, etc., the equipment can be connected to the central computer. The CPU is the focus of the computer system, as illustrated in Fig. 9.2.

The sets of instructions for the CPU are stored in RAM. These stored programs give the computer a group of operations to perform. Aside from storing program instructions, the main memory usually holds the data being worked on. Main memory can be visualized as a group of mailboxes, each having a unique location number or address. When the information stored in the box is required, the computer goes to that location and collects the required data. Main memory is expensive, so secondary memory is used for data storage when practical.

Data is usually stored and processed in units of words, which usually consist of 8, 16, or 32 binary digits. The usual word lengths for typical computers are presented in Table 9.1. The larger the word size, the more information may be processed at once, and accordingly the operation will proceed at a faster pace.

9.2 Computer-Aided Design and Drafting

Initially, CADD evolved as a means of speeding the production of two-dimensional line drawings. From this early assignment the computer's duties have been expanded to include the generation of three-dimensional drawings, pipe and cable tray routing, interference checks, and the coordination of related documents, such as piping drawings, stress isometrics, and erection and fabrication drawings.

Today, most CADD systems can draw a view from a distance, a profile, a plan; zoom in on any portion of a drawing; or generate three-dimensional images. Thus the computer-aided designer can see the system from numerous views which increases the ability to visualize the piping system and area of installation. This ability to view from different points of perspective is a design aid that is particularly useful for reviewing space available for maintenance and installation problems. The ability to interact with the design gives added flexibility to make quick changes or many editions of the same design with minor changes.

With standard drafting symbols stored in the computer memory, the designer can draft much more quickly than a manual drafter and in a manner that is easier to read and standardized. Structural details, piping, in-line equipment symbols, and weld symbols all comprise a menu to which the designer has immediate access.

The CADD system hardware may be comprised of one or more input terminals that serve as design work stations: the computer, plotting equipment, and the communications network linking them. Such a system is shown in Fig. 9.4. This system is usually purchased with the user's day-to-day needs, as well as future options, in mind.

Since the CAD/CAE industry is rather young, with many manufacturers producing components of the complete system, the problem of compatibility must be considered. Equipment of one manufacturer may not be fully compatible with all equipment of other manufacturers. For example, there are two major approaches to generating images, known as *stroke* (or *vector*) *writing* and *raster*. A stroke-writing device draws lines to create an image, while a raster device uses an array of closely spaced dots (known as *pixels*) to form a picture. Stroke-writing devices are commonly divided into two types: storage tube and refreshed displays. Storage tubes are usually monochromatic with a resolution meeting or exceeding most

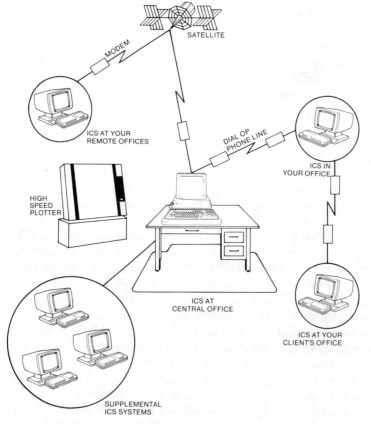

Figure 9.4 Typical integrated CADD system. *(Courtesy of Engineering Planning and Management, Inc.)*

other devices, but drawing speeds are relatively slow—typically 5000 in/s (12,700 cm/s). A vector refreshed display can provide dynamic rotation and viewing capabilities. It is usually faster than the storage tube device, typically 800,000 in/s (2,032,000 cm/s).

A raster screen consists of pixels laid out on a series of horizontal lines. Each pixel is a color combination of simultaneously displayed colored dots using primarily red, green, and blue. These pixels can be turned on or off, with different colors, to make up the total picture. The raster display also contains an image storage system within RAM.

Software packages and data storage formats may also not be compatible with all manufacturers' systems because no one has clearly been given the lead by the industry. Therefore, the equipment selection should be made after careful consideration of future plans and current needs.

The CADD hardware of the 1980s makes the designer's work station look like a far cry from the traditional drafting board. In front of the designer are one or more televisionlike screens. The screens are called CRTs. The screen provides an image that satisfies the need of the designer to see the design progress. The screens shown in Figs. 9.4 and 9.5 are connected by a short cord to a keyboard. The keyboard is used to enter the data from which the software generates the image on the screen. The CADD system shown in Fig. 9.5 is also controlled by a menu pad.

The menu pad does not have the traditional tools of manual drafting such as T-squares, rulers, pencils, triangles, etc. Instead, the designer uses a type of electronic pen to select one of the symbols shown on the menu pad. Once the pen is positioned over the desired symbol and the correct command is given, the symbol appears instantly on the screen and can be moved by the designer to any location. The pen can provide a series of these symbols at any location on the screen within seconds. Additional symbols can be added until the information required for a complete drawing has been entered. Other commands can be used from the menu to provide standardized items such as overhead steel, walls, duct work, lighting, and floors and ceilings. The symbols for each detail can be positioned anywhere on the screen and labeled as desired. Color schemes may be varied in order to permit numerous systems (piping, raceways, HVAC, etc.) to be displayed simultaneously while maintaining easy identification. Once the design has been completed, hard copies can be made of the drawing, its mirror image, or any variations. Additionally, a finished document, such as a piping drawing, could be easily used as a basis for generating a second document, such as a stress isometric. The delete and add features of the menu pad can be quickly used to change the image on the screen until the drawing is exactly as desired.

With this equipment available to architects and engineers, a complete building, inside and out, can now be viewed before a shovel is pushed into the ground.

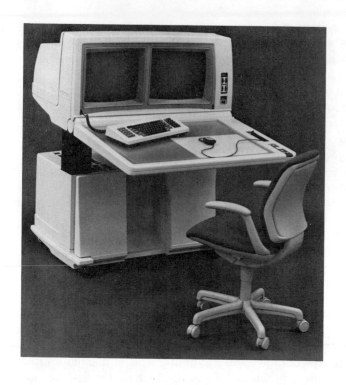

Figure 9.5 (*a*) Typical turnkey CADD system; (*b*) closeup of typical turnkey CADD system. *(Courtesy of Intergraph Corporation.)*

To be most effective, all the related documents, such as the flow diagram, piping drawing, and isometrics, should be on the same data base, thus minimizing the data entry time and ensuring compatibility among the documents. The sharing of the data and system permits the modular development of software and results in a standardization of procedures and symbols used on the project and within the company.

Once all design information has been entered into the data base, the computer has the capability to perform other tasks, such as interference checks, verifying the size of equipment for shipping, or even locating equipment or routing piping that must be added. It is a fairly easy matter for the computer to compile a bill of materials from the data base. Composite drawings, such as that shown in Fig. 7.2b, can be generated to assist designers in all disciplines. The CAD process has evolved to a stage where virtually all information required for engineering design can be computerized. Figure 9.6 illustrates how all documents on the project may be maintained and generated through the use of a single CADD system.

The physical size of each drawing may be varied based on the information contained on it. The sizes of the production drawings are usually standardized by the decision of the engineering department head. Some organizations prefer the smaller drawings because of ease of handling and filing, while others prefer larger drawings.

9.3 Computer-Aided Engineering

9.3.1 Introduction

In the 1960s computers were introduced to piping design on a large scale. The use of computers has grown to a point where the term might better be computer-*dominated* engineering rather than computer-aided engineering (CAE). Originally, computerized engineering calculations were performed through access to a large computer system called a *mainframe*. Mainframes have large storage capacities which enabled the computer to perform sophisticated matrix manipulations that manually were so time-consuming as to be impractical.

The mainframe capacity was so large that it could perform stress analysis calculations that were not considered possible just earlier. One of the earliest, and most widely used, CAE applications was the integrated civil engineering system (ICES) developed at the Massachusetts Institute of Technology. This series of programs covered a wide range of applications in the civil engineering industry. The most prominent program in the ICES library is called STRUDL, which stands for *stru*ctural *d*esign *l*anguage. STRUDL was one of the first structural engineering programs used nationally, and it still maintains great popularity. An in-depth discussion

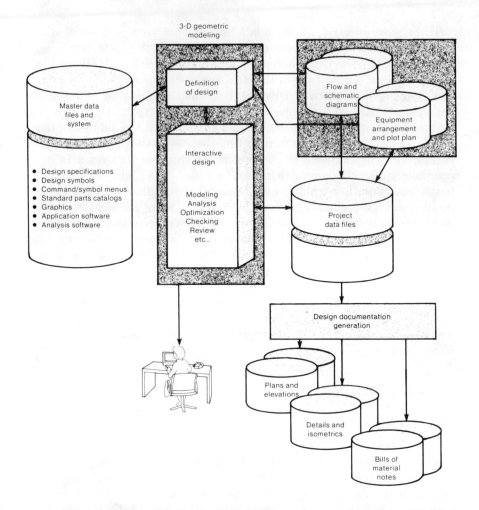

Computer-Aided Design Engineering

Figure 9.6 Interaction of CADD system with project documents. *(Courtesy of Engineering Planning and Management, Inc.)*

of the STRUDL program as it relates to pipe support design appears later in this chapter. A sample pipe stress computer analysis is also presented later in this chapter.

All structural programs require a similar set of data that defines the characteristics of the system. First, the geometry of the structure, such as number and lengths of members and location of member junctions, must be entered. Second, the member properties, such as member cross-sectional and material parameters, must be described. Third, the degree of support, such as system support points and internal joint releases, must be defined. Finally the loads acting on the structure must be entered. Based on this input, the program will be able to calculate the response of the system in terms of support loads, internal loads, deflections, and stresses. Data preparation for a typical structural program, as well as a sample computer problem, is described later in order to reinforce the information presented here.

Once the piping layout or support structure has been determined, the computer model can be prepared. A computer can perform an analysis only if it has a mathematical model or numerical description of the system on which to operate.

A computer model is made by dividing the structure into segments, called *elements,* superimposed on a coordinate grid system. Node points or joints define the location of the elements and must be called out at all points of interest, such as element intersections, points of loading, concentrated-mass locations, or simply locations where stress or deflection results are desired. The elements represent the stiffness of the connections between the joints. The loaded system is considered as being built up of numerous interconnected substructures, or elements. Thus the words *finite element* are used to describe the process. Since these elements can have almost any stiffness parameters and can be put together in virtually any orientation, they can describe objects exceptionally well.

The finite-element method had its birth in the aerospace industry in the early 1950s. The first publication on this subject showed the application of the finite-element method to heat transfer applications. Later research showed that the finite-element method could also be applied to the field of structural mechanics.

Finite-element analysis requires the solution of a set of simultaneous linear equations. In precomputer days the large amount of tedious calculations involved in this task posed a limitation on the applicable theory. Theories such as moment distribution were applied as manual calculation estimates of the magnitude of loading. The introduction of high-speed digital computers made finite-element analysis more feasible for industrial applications.

The finite-element method of structural analysis has several advantages. Some of these are as follows:

1. The material properties or element cross sections of adjacent elements need not be the same. This allows complex structures made of several materials and member types to be modeled.
2. Irregular shapes can be approximated by using straight-sided elements, making it possible to model complex structures and boundaries.
3. The size of the elements can vary. Thus the grid size can expand to fit the complexity of the problem.

4. Numerous boundary conditions can be used, including mixed types of boundary conditions.
5. All the above conditions can be incorporated into one general computer program. The limiting factors are usually the computer memory size and computer cost when complex problems are solved.

The simplest type of element used in structural analysis is the single-degree-of-freedom element. This element is free to distort in one direction only (normally axially) and therefore can only take a load in a single direction. Common examples of one-dimensional elements are pipe supports in a pipe stress analysis problem and truss elements in a structure. The geometry of this type of element is defined by two node points (start and end) or simply by a direction of action and a single node point of application.

Multidegree-of-freedom elements are more common in most structural problems. They are free to distort in more than one degree of freedom. Most real-life members are free to distort in six degrees of freedom (translation and rotation about three orthogonal axes). The multidegree-of-free-

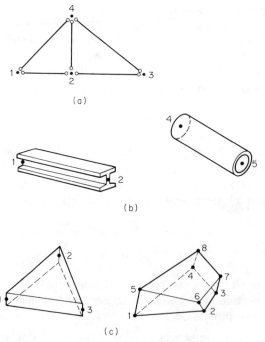

Figure 9.7 Types of finite elements: (*a*) one-dimensional, one-degree-of-freedom (truss) elements; (*b*) one-dimensional, multidegree-of-freedom elements; (*c*) multidimensional, multidegree-of-freedom elements.

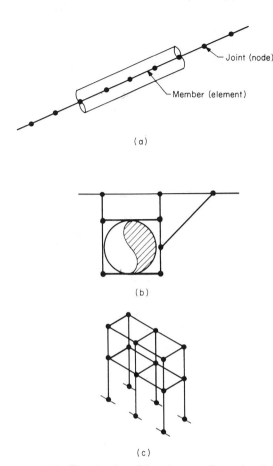

Figure 9.8 Structural models using one-dimensional elements: (*a*) one-dimensional model; (*b*) two-dimensional model; (*c*) three-dimensional model.

dom element may be defined by two node points (along the element axis) along with cross-sectional properties or by corner node points for irregularly shaped elements. Examples of various types of elements are shown in Fig. 9.7. Multidimensional elements are convenient for describing unusually shaped structures.

Most pipe stress and pipe support structural computations use one-dimensional, multidegree-of-freedom elements. Figure 9.8 shows that one-dimensional elements can be used to describe one-, two-, or three-dimensional structures.

Support structures are usually composed of a series of individual members which are connected at their ends. The member represents the geometric center or the center of curvature superimposed on a coordinate grid system. The members intersect at points called *joints,* or *node points.* These joints are usually numbered for identification purposes and located by their cartesian coordinates.

Simple support calculations can be analyzed by manual methods such as those presented in Chap. 8. More complex structures can be solved by considering the behavior of each individual element separately and assembling the results in such a way that at each node point the following rules are satisfied:

1. Equilibrium of forces and moments
2. Laws of material behavior
3. Compatibility of displacements and rotations

Multiple-member structures, such as many support structures, can be very laborious to calculate. Each joint requires a separate equation for each degree of freedom of distortion. Thus even simple structure calculations require a large number of simultaneous equations (for example, a structure with 10 joints in three dimensions—six degrees of freedom—requires 60 equations). Since more than a few simultaneous equations are extremely time-consuming to solve, the need for a computer becomes apparent even for simple problems.

The use of computers allows the engineer to solve a large number of simultaneous equations with speed and accuracy. The matrix method used for structural analysis is usually one of the following:

1. Stiffness (displacement-related)
2. Flexibility (force-related)

The stiffness method appears to be more popular because of certain drawbacks of the flexibility method (inadequate ability to handle free ends and internal loops, for example).

Figure 9.9, a flowchart for the stiffness method of structural analysis, shows the steps a typical program performs during its evaluation. The computer program takes input data which must fully describe the problem. Local stiffness matrices for each element are generated. These are assembled to form the stiffness matrix for the entire structure. The matrix is then modified to reflect boundary (restraint) conditions. From the stiffness matrix K and the force vector F, displacements at each joint are found. By multiplying the relative joint displacements by the local stiffness matrices, the internal member loads can be found and applied to the cross-sectional properties to find member stresses. The complete results (joint displacements, member and restraint forces, and member stresses) are then printed out.

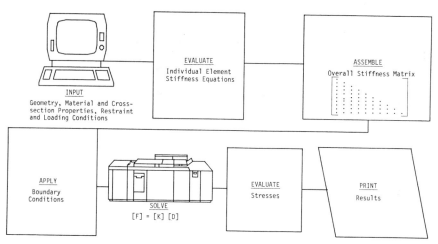

Figure 9.9 Flowchart for stiffness method of structural analysis. *(Courtesy of Impell Corporation.)*

9.3.2 Input methods for computer-aided engineering

The support engineer will most often encounter two major types of CAE programs: the structural analysis program and the pipe stress program. The structural analysis program is exactly what it seems—a program for analyzing the response under loading of a system made up of structural members. This type of program will normally have input methods which facilitate the use of common structural shapes, typical connection conditions (pinned or fixed joints), and common loading conditions. The most prominent of these programs is the ICES STRUDL program.

Pipe stress analysis programs are no more than very specific versions of structural programs, with input methods which facilitate the modeling of piping systems. The beam elements have pipe cross sections, the connections are continuous (fixed), support conditions can be easily modeled to simulate actual support types, and loading conditions lean toward those associated with piping (deadweight, thermal, pressure, seismic, etc.). The relative capabilities of some pipe stress programs available on the market are shown in Table 9.2.

Since both types of programs are inherently similar, and since all structural and piping programs require the same basic information, the required input for only one program needs to be presented here. STRUDL will be used as the model program. An annotated sample pipe stress analysis is included later in the chapter for reference.

Most computer program input consists of command words and data

TABLE 9.2 Pipe Stress Capabilities

Program Capability	ADLPIPE	NUPIPE	PIPESD	TRI-FLEX	I-PIPE	SUPER-PIPE
Design Codes:						
ASME, Sec III, Class 1	●	●	●			●
ASME, Sec III, Class 2	●	●	●	●	●	●
ANSI B31.1	●	●	●	●	●	●
USAS B31.1	●	●				●
ANSI B31.3	●	●	●	●		●
ANSI B31.4	●			●		●
ANSI B31.8				●		●
NEMA				●		
API-610				●		
API-617				●		
ASME, Sec III, NB-3647				●		
Analytical Capabilities:						
Pressure	●	●	●	●	●	●
Temperature	●	●	●	●	●	●
Deadweight	●	●	●	●	●	●
Seismic (dynamic)	●	●	●			●
External forces	●	●	●	●	●	●
Time history	●	●	●			●
Independent support motion	●	●	●			●
Fatigue	●	●	●			
Thermal transient	●	●	●			●
Pipe rupture		●				●
NRC Guide 1.92	●	●	●			●
Components:						
Straight pipe	●	●	●	●	●	●
Elbow	●	●	●	●	●	●
Miter	●	●		●		
Tee	●	●	●	●	●	●
Branch	●	●	●	●	●	●
Snubber	●	●	●	●	●	●
Reducer	●	●	●	●	●	●
Valve	●	●	●	●	●	●
Beam	●	●	●	●		●
Bellows		●				
6 × 6 Matrix	●	●		●	●	●
Nonlinear gap		●				
Nonlinear restraint		●		●		
Spring	●	●	●	●	●	●
Units:						
USCS	●	●	●	●	●	●
Metric	●	●	●		●	●
SI	●	●	●	●	●	●
Features:						
Interactive input					●	
Free format input	●	●	●	●		
Plotting	●	●	●	●	●	
Automatic mass lumping	●	●	●			●
Code compliance report	●	●	●	●	●	●
Automatic stress factor	●	●	●	●	●	●
User stress factor	●	●	●	●	●	●

SOURCE: From *Piping Products Capabilities Summary,* Scientific Information Services of Control Data Corporation. Reprinted by permission. Additional information courtesy of Impell Corporation.

stored as records on punched cards, disk, or tape. The data lines for a particular computer input are given either as a group of lines typed on a terminal or as a deck of punched cards containing the coded program. Thus the words *line* and *card* are used interchangeably in discussing data input.

Usually one command is allowed on each card line. Sometimes a special character, such as semicolon, can be used to separate the commands. Although this would be equivalent to ending an input record, it is not recommended for use since it may cause confusion. There are 80 columns on a card and up to 132 columns per line on an interactive terminal, with each column being the location of a character of input data. Data may be continued to the next record if a special symbol (such as a minus sign preceded and followed by a space) is given as the last character on the line. Usually the number of continuation cards is limited. Some programs allow a command or data entry to be placed anywhere on a line; this is called *freefield format*. However, some other commonly used programs have a definite location for certain or all input data. This type of input is called *fixed format*. Certain data lines may be used for comments when a predetermined symbol, such as a dollar sign ($) appears on the line. Any data following this sign is nonexecutable data and is only reprinted. The exact program of interest should be reviewed to see what, if any, special symbols are used. Input data may be required in the form of integers, real numbers, or alphanumeric characters. Integers are whole numbers without decimal points and are usually used for labeling, such as with member or joint numbers. Decimal numbers are real numbers and are required for physical data such as geometric, cross-sectional, material, or loading properties. Alphanumeric input may include both letters and numbers and is most often used for naming purposes. Alphanumeric input is often delimited by single quotation marks.

Most structural analysis program user manuals indicate the commands required to perform the procedure desired. Many use common symbols to describe the options available with each command. The option symbols used in the STRUDL user manual (as published by M.I.T.; different versions may have enhancements not discussed here) are typical. They are as follows:

1. *Parentheses.* ( ) mean that commands contained within are optional and, if omitted, would not affect the executability of the command.

2. *Braces.* { } mean that a choice exists between the enclosed options, of which only one choice is possible and must be made.

3. *Asterisk.* * means that one or more choices of command options are available. The choices can appear in any order.

4. *Brackets.* [ ] mean the data labels within the brackets may be omitted when the data items are entered. When labels are used, the data items may be entered in any order. If no labels are given to the data, it is assumed that the information entered corresponds to the order shown in the user manual.

5. *Arrow.* → means that the option indicated is assumed as the default if no choice is indicated by the user.

$$\left\{ \begin{array}{c} \to X \\ Y \\ Z \end{array} \right\}$$

The above means that if no label is given to the data, it is assumed to conform to the X variable.

6. *Underline.* ___ means that the underlined portion of the command is the minimum initial characters that the computer needs to recognize that word. The remainder of the command need not be written out.

7. *Letters.* Uppercase, or capital, letters are usually used to indicate command words. Lowercase letters (i, v, and a) usually indicate input data that must be supplied by the user, in integer, real, and alphanumeric forms, respectively.

Before the system is described to the structural program, the system of units must be agreed upon. The default units for STRUDL input and output are length = inches, force = pounds, angle = radians, temperature = Fahrenheit, and time = seconds. These units may be overridden by using any of the units shown below:

Units	Length	Force	Angle	Temperature	Time
Default	INCHES	POUNDS LBS	RADIANS	FAHRENHEIT	SECONDS
Alternates	FEET FT CENTIMETERS CMS METERS	KIPS TONS MTONS KILOGRAMS KGS	DEGREES	CENTIGRADE	MINUTES HOURS

To change the input or output units used in the STRUDL program, the UNITS command is used, identifying all units used in the anlaysis or only those changed:

UNITS   INCH   POUNDS   DEGREES   SECOND

Next the type of structure being analyzed must be described. First, the classification of the problem as two-dimensional (plane) or three-dimensional (space) must be made. This classification determines how many axes will be used to describe the problem.

Next the types of elements used must be identified by classifying the model as a frame, grid, or truss, as shown in Fig. 9.10. Frames and grids are structures in which all joints are initially assumed to be rigidly connected to the members unless a specified release-of-rigidity command is used in the program. The frame uses primarily beam elements while the grid uses primarily multidimension finite elements. A truss is a structure in which all joints are assumed to be pin-connected. The truss members take only axial forces, and the support point reactions are only forces, not moments. The types of structures available in the STRUDL program are listed below:

$$\underline{\text{TYPES}} \text{ (of structure)} \quad \underline{\text{PLANE}} \begin{Bmatrix} \text{FRAME} \\ \text{GRID} \\ \text{TRUSS} \end{Bmatrix} \begin{Bmatrix} \rightarrow \underline{\text{XZ}} \\ \underline{\text{YZ}} \\ \underline{\text{XY}} \end{Bmatrix}$$

$$\underline{\text{SPACE}} \begin{Bmatrix} \text{FRAME} \\ \text{TRUSS} \end{Bmatrix}$$

The first step in describing the structure to the computer is to draw a model of the system. Joints should be used at the intersections of members, at points of loading, at points of support, and at other points of interest. Members run between the joints, conforming to the longitudinal axes of the real-life structural members. All members are considered to be straight, prismatic, and having constant material properties.

The numbering sequence for joints and members should be selected with consideration to the ease of reviewing the input or output data. The numbering of a system should be done in a systematic and logical manner. Numbering the joints so as to minimize the difference between joint numbers at opposite ends of members usually reduces what is called the *bandwidth* and results in a more efficient and less costly program execution time. To clarify this discussion about bandwidth, the reader is referred to Fig. 9.11. By numbering the joints vertically in this example, the resulting bandwidth is 6. By numbering the same structure horizontally, the bandwidth is reduced to 4, a 33 percent reduction. Since the computer cost increases linearly with the bandwidth, it may be important to number the joints carefully. However, most structural programs have an option which automatically internally renumbers the joints to minimize the bandwidth. Therefore this is usually not a concern. Member numbering is not as critical as joint numbering, yet an intelligent numbering system can reduce the time spent reviewing input and output data.

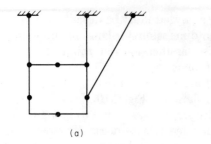
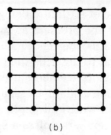
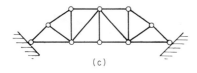
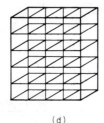

Figure 9.10 Types of structures: (*a*) rigid frame, two-dimensional, all joints fixed; (*b*) grid or mesh, two-dimensional; (*c*) truss, two-dimensional, all joints pinned; (*d*) grid or mesh, three-dimensional.

Performing structural analysis requires the use of two coordinate systems. One coordinate system locates the joints of the structures and is called the *global coordinate system*. The second coordinate system is called the *local coordinate system* and identifies the member longitudinal x axis as well as the y and z principal axes of the cross section.

The global coordinate system has three mutually orthogonal reference axes, with which any location on a structure may be defined. The global coordinate system is used to locate points on the model (joints) as well as define loading directions. The location of X = 0, Y = 0, and Z = 0 is called the *origin* of the mathematical model and is the point of reference from which the structural model is oriented. The origin can be located arbitrarily on the model, since all locations are relative.

The geometry of the structure is first described by entering the numbers and locations of the joints. The means of locating the joints on the mathematical model is through the use of X, Y, Z coordinates with reference to the point of origin. All dimensional data used in defining the model geometry is with reference to this position. The global coordinate system is called the *cartesian coordinate system*. The axes of the coor-

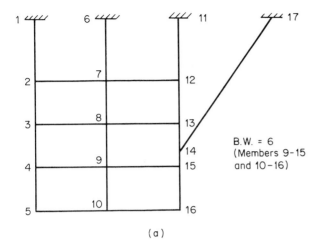

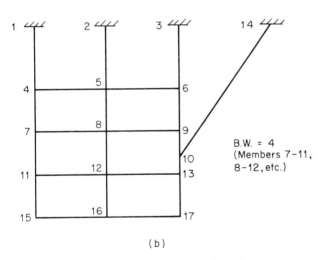

Figure 9.11 Joint numbering for improved bandwidth: (*a*) vertical; (*b*) horizontal.

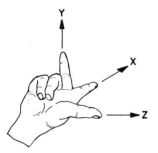

Figure 9.12 Right-hand rule for axis orientation.

dinate system may be identified by using the right-hand rule shown in Fig. 9.12.

The general form of the joint coordinates command is

$$\begin{Bmatrix} \underline{\text{JOINT}} \\ \underline{\text{NODE}} \end{Bmatrix} \begin{Bmatrix} i_n \\ "a_n" \end{Bmatrix} \underline{\text{COORDINATES}} \begin{Bmatrix} \rightarrow \underline{\text{FREE}} \\ \underline{\text{SUPPORT}} \end{Bmatrix}$$
$$([\underline{X}\text{COORD}] \ v_x \ [\underline{Y}\text{COORD}] \ v_y \ [\underline{Z}\text{COORD}] \ v_z)$$

The command JOINT COORDINATES need not be repeated with every entry, but may be used as a heading for all the joints entered. The lines would be on the form

$$\begin{Bmatrix} i_n \\ ,a_n, \end{Bmatrix} ([\underline{X}\text{COORD}] \ v_x \ [\underline{Y}\text{COORD}] \ v_y \ [\underline{Z}\text{COORD}] \ v_z) \begin{Bmatrix} \rightarrow \underline{\text{FREE}} \\ \underline{\text{SUPPORT}} \end{Bmatrix}$$

Each of the lines of data contains the coordinates of a joint in the model until all joints have been entered. For tracking purposes, the joints should be listed sequentially. For the model shown in Fig. 9.13, the joint coordinates are listed below in the format described above:

		JOINT COORDINATES		
$ Joint no.	X coordinate	Y coordinate	Z coordinate	Support
1	0.0	0.0	0.0	S
2	10.0	0.0	0.0	S
3	0.0	−10.0	0.0	
4	5.0	−10.0	0.0	
5	10.0	−10.0	0.0	
6	10.0	−12.5	0.0	
7	0.0	−15.0	0.0	
8	10.0	−15.0	0.0	
9	20.0	0.0	0.0	S
10	0.0	−20.0	0.0	
11	5.0	−20.0	0.0	
12	10.0	−20.0	0.0	

Note that the comment line following the dollar sign is not executable and is included only for clarity.

An S at the end of a joint entry line indicates that the joint serves as a restraint (support) point. The boundary condition at the restraint—whether fixed or pinned—is an important consideration that must be reviewed for each structure and its corresponding model. The joint release command may be used to release rigid joints when a restraint is not fully restrictive in all degrees of freedom. This command is most often used to release moments at pinned connections, but may also be used for releasing forces.

The typical format of the joint release command is

$$\left\{ \begin{array}{l} \underline{\text{JOINT}} \\ \underline{\text{NODE}} \end{array} \right\} \underline{\text{RELEASES}} \text{ (JOINT LIST)} * \left\{ \begin{array}{l} \underline{\text{FORCE}} \\ \underline{\text{MOMENT}} \end{array} \right\} * \left\{ \begin{array}{l} X \\ Y \\ Z \end{array} \right\}$$

or

```
JOINT RELEASE
$ Joint numbers Force X, Y, Z moment X, Y, Z
  1 2     MOMENT Y Z
  9       FORCE X     MOM Z
```

After the joints have been entered, the members must be described. Each member must be labeled and defined by identifying the joint at the beginning and the end. The member incidence data usually begins with a line that labels the set of data to follow. The member incidences for the model

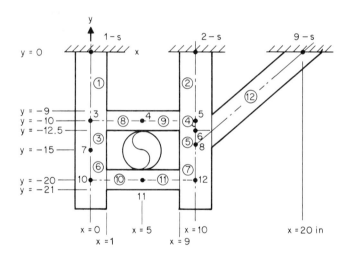

Figure 9.13 Structural model.

shown in Fig. 9.13 are presented here:

	MEMBER INCIDENCE	
$ Member No.	Start Jt.	End Jt.
1	3	1
2	5	2
3	7	3
4	6	5
5	8	6
6	10	7
7	12	8
8	3	4
9	4	5
10	10	11
11	11	12
12	8	9

Later in the program it may be necessary to refer to a particular end of a member; in this case the member "start" or "end" will be referenced. Remember that the member "start" and "end" points are as identified in the member incidence command.

Much as the rigidity of the boundary conditions could be modified through the joint release command, the rigidity of the internal joints may be modified by using the MEMBER RELEASE command. This command releases degrees of freedom (forces and moments) at the connections between members. The releases are given in relation to the member local coordinates, where the X axis corresponds to the longitudinal axis of the member and the Y and Z axes correspond to the weak and strong axes of the cross section, respectively (as shown in Fig. 9.14). The format for this command is as follows:

$$\text{Member list} \left\{ * \left[\begin{array}{l} * \left| \text{START} \right| \left\{ * \left[\begin{array}{l} \text{FORCE} \\ \text{MOMENT} \end{array} \right] * \left\{ \begin{array}{l} X \\ Y \\ Z \end{array} \right\} \right\} \\ * \left| \text{END} \right| \left\{ * \left[\begin{array}{l} \text{FORCE} \\ \text{MOMENT} \end{array} \right] * \left\{ \begin{array}{l} X \\ Y \\ Z \end{array} \right\} \right\} \end{array} \right] \right\}$$

MEMBER RELEASE

An example of the written command is

```
         MEM REL
5 STA FOR X Y    END FOR Y MOM Y
7 STA FOR X
```

The MEMBER END JOINT SIZE command corrects the model for any incorrect member lengths due to the specification of the joint locations at the center of the attached members. Figure 9.15 shows some

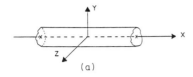

(In Figures below X is out of paper)

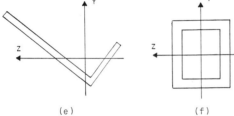

Figure 9.14 Member local coordinates: (a) local principal axes orientation; (b) W shapes and I beams; (c) tees; (d) channels; (e) single angles; (f) tubes and boxes.

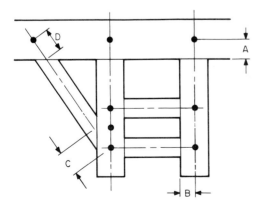

Figure 9.15 Member end joint sizes.

examples of locations (labeled A, B, C, and D) where the MEMBER END JOINT SIZE command could be used to correct member lengths. This command adjusts the member stiffness matrix to account for the effectively rigid portion of length and is therefore more useful in cases where the end joint size is relatively large in proportion to the total member length. The applicable command form is

```
MEMBER END (JOINT) SIZE
Member list [START ] V₁ [END ] V₂
```

When the MEMBER END JOINT SIZE command is used, members (referring to Fig. 9.13) with the same size and location of end joint may be grouped as shown:

```
MEMBER END JOINT SIZES
$ Member Number, Location, Joint Length
8 10    START 1
9 11    END 1
12      START 1.414
```

The MEMBER ECCENTRICITY command may be used to modify member geometry under certain circumstances. Before discussing the form of the input command, let us discuss an eccentric load in general. Referring to Fig. 9.16a, we see the vertical load F_y is normal to the member, and thus its load is transmitted along the line of action to the centerline of the member. The F_x load is applied eccentrically to the structural member, causing a moment equal to the load multiplied by the distance from the point of load to the support member centerline. One method of modeling this situation is shown in Fig. 9.16b—that is, to locate joint 2 at the point of loading. This ensures that the eccentric loading on the structure as a whole will be considered correctly. However, this model incorrectly represents the centerline of the members as not parallel to the x axis. This inaccuracy can be corrected through use of the MEMBER ECCENTRICITY command, which tells the program that the centerline of the member does not run through one or both endpoints.

When member eccentricities are identified, it is necessary to identify the member number, the end of the member having the eccentricity, the direction of the eccentricity, and the distance of eccentricity. As in most input commands, a command identifier is used before the data lines are input. An example of the format of a MEMBER ECCENTRICITY command is:

```
MEMBER ECCENTRICITIES ( |GLOBAL| )
                      ( |LOCAL | )
Member list *  [START  [X] V₁ [Y] V₂ [Z] V₃]
               [END    [X] V₄ [Y] V₅ [Z] V₆]
```

or

```
      MEM ECC
$ Member Number  Eccentricity  Location  Direction  Amount
        1                        END        Y        -6.5
        2                        START      Y        -6.5
```

Once the geometry of the structure has been entered, it may be verified visually through the PLOT PLANE command. Two typical methods are normally used for generating plots of the model. One method is to have the plot printed by the printer as part of the output data, and the second method is to have a separate plotter draw the model. The plotter drawing will be a more accurate representation of the model. However, since the purpose of the plot is only to confirm the general shape of the model, the printed plot is usually adequate. The plot can be made of any plane (parallel to two global axes) cutting through the model. The plot will show all joints and members falling within that plane. The plane to be plotted may be specified by listing two axes and one joint, or three noncollinear joints, as shown in:

```
PLOT PLANE XY 10
PLOT PLANE    7 10 12
```

As many plots as are desired may be called for in order to view the model from all sides and at many section cuts.

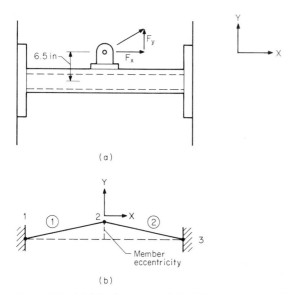

Figure 9.16 (*a*) Member eccentricity; (*b*) geometric model.

Once the model geometry has been entered and verified, the member cross-sectional and material properties must be entered. In most cases the members will be prismatic, which means that their properties do not vary with length. Therefore, when member properties change, the number of the member should change also.

Material properties that are required for each member are listed in Table 9.3. When a group of members has the same properties, these may be grouped under a single input line. Various methods of inputting member constants are available, as shown by the command format below. Units are the same as those specified previously:

$$*\begin{Bmatrix} \underline{E} \\ \underline{G} \\ \underline{CTE} \\ \underline{DENSITY} \\ \underline{BETA} \\ \underline{POISSON} \end{Bmatrix} \begin{Bmatrix} V_1 \text{ list}, V_2 \text{ list}, V_3 \text{ list, etc.} \\ V_1 \underline{ALL} \\ V_1 \underline{ALL\ BUT}\ V_2 \text{ list}, V_3 \text{ list, etc.} \end{Bmatrix}$$

The only member constant shown in Table 9.3 that does not indicate a material property is BETA. This constant represents the orientation of the member to the standard member orientation (BETA = 0°). The BETA angle measures the angle of rotation (with clockwise positive) of the cross section about the local X axis, with the standard orientation as a reference.

For STRUDL, the standard orientation is defined as follows:

1. For members running in the horizontal plane, the local Y axis is parellel to the global Y axis.

2. For members running parallel to the global Y axis, the local Z axis is parallel to the global Z axis.

3. For skewed members, the projection of the global Y axis onto the local YZ plane is parallel to the local Y axis.

Figures 9.17 and 9.18 demonstrate BETA for a member running in the horizontal plane and for a skewed member.

TABLE 9.3 Member Constants

Command	Description	Units
E	Modulus of elasticity	Force/length^2
G	Shear modulus	Force/length^2
DENSITY	Density	Weight/length^3
POISSON	Poisson ratio	Dimensionless
CTE	Coefficient of thermal expansion	Length/length/temperature
BETA	Orientation of member about local X axis	Angle of rotation

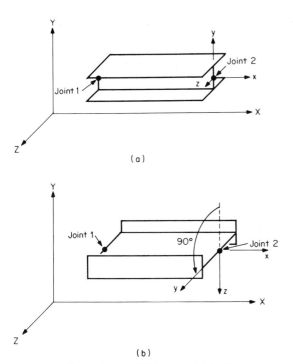

Figure 9.17 Examples of BETA angle: (*a*) BETA = 0°; (*b*) BETA = 90°.

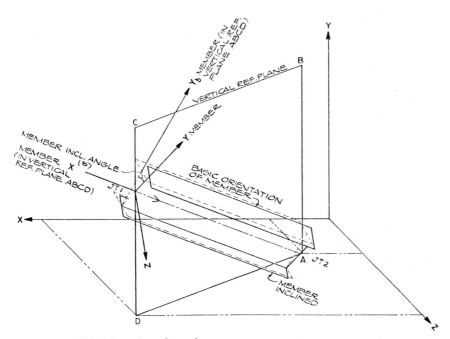

Figure 9.18 BETA for a skewed member.

305

Next, the member cross-sectional properties must be given. These properties include, among others, the cross-sectional area A_X; shear areas A_Y and A_Z; moments of inertia about the three axes I_X, I_Y, I_Z; section modulus S_Y, S_Z; and radius of gyration R. These properties may be input to the program or may be extracted from tables which contain the properties for standard structural shapes such as wide flanges, angles, tubes, etc.

MEMBER PROPERTIES commands take the following form:

```
MEMBER PROPERTIES          ⎧ PRISMATIC (Section values)             ⎫
                           ⎪ TABLE 'Table name' ('Section name')    ⎪
                  (list) ⎨ VARIABLE (Section values)               ⎬
                           ⎪ FLEXIBILITY (Matrix specs)             ⎪
                           ⎩ STIFFNESS (Matrix specs)               ⎭
```

For example,

```
MEM PRO PRI
1 to 11 13 15 AX 15.6 IX 0.98 IY 425. IZ 95.8
```

or

```
MEM PRO TAB 'STEEL W78'
1 TO 11 13 15 'W12x53'
```

The MEMBER PROPERTIES command applies to all members that are listed after the initial title command. Members with the same properties can have them identified with a common-input command statement. The units of the member properties are those specified previously in the program.

At this point the entire structure has been described in terms of geometry, boundary conditions, material properties, and member cross sections. From this information the local and global stiffness matrices can be assembled. The only data still required to describe the problem is the loading conditions.

A structure restraining a number of pipes, each with a number of loading conditions (such as normal, upset, emergency, and faulted) could result in a large number of loading combinations. This could be very expensive to run and might provide data of marginal value. Thus, in selecting the loading cases care should be exercised so only probable worst-case loadings are used in the analysis.

The commands for static loading in STRUDL are in the form of (1) joint loadings, (2) member loadings, (3) joint displacements, or (4) member temperature loading. The LOAD command is used to identify the loading conditions occurring in a given load set. Each loading case usually is identified by a number and may also be given a name. The input commands usually must identify the member or joint in the model to which the loading is applied, the type of load, the direction, and the magnitude. A DEAD LOAD command is usually also available which finds the self weight of each member and applies it to the structure as well.

When there are a number of independent loadings, a sequence of loading conditions can be analyzed separately as shown:

```
UNITS INCH POUNDS
LOAD 1 'UPSET 1'
JOINT LOADS 1 FORCE Y 8000.00 2 MOMENT Z 6000.
MEMBER LOADS 1 4 FORCE Y UNIFORM W −2.0
DEAD LOAD − Y

LOAD 2 'UPSET 2'
JOINT LOADS 1 FORCE Y −8000.00 2 MOMENT Z −6000.
MEMBER LOADS 1 4 FORCE Y UNIFORM W +2.0
DEAD LOAD −Y
```

Forces and moments applied to joints may be specified through the JOINT LOADS command. The format of the command is

$$\underline{\text{JOINT LOADS}} \; * \; \begin{Bmatrix} \underline{\text{F}}\text{ORCE} \; [X] \; v_1 \; [Y] \; v_2 \; [Z] \; v_3 \\ \underline{\text{M}}\text{OMENT} \; [X] \; v_4 \; [Y] \; v_5 \; [Z] \; v_6 \end{Bmatrix}$$

Use of the JOINT LOADS command can be seen in the above examples.

The JOINT DISPLACEMENTS command can account for imposed displacements due to factors such as postconstruction foundation settlement, seismic anchor displacements, and pipe thermal movements. The JOINT DISPLACEMENTS command operates in precisely the same way as the JOINT LOADS command, previously described, except that the values specified are joint displacements applied to support joints in restrained directions. A sample JOINT DISPLACEMENTS command is

```
LOAD 5 'ANCHOR MOVEMENTS'
JOINT DISPLACEMENTS

2 DISP X 1.8 Y −1.3
3 DISP X 2.0 Y −1.8
```

The format assumes that the applied forces, moments, and displacements are parallel to either the X, Y, or Z axis. If the loadings are not parallel to one of the three major axes, the loads must be broken into components which are parallel to these axes.

A number of different member loads may be applied to a member. Member loads can be in the form of member force or member moment loads. The loads can vary along the length, be constant along the length, or be concentrated at specified locations. The member loads can be specified in the local or global coordinate systems, with the local usually being the program default.

The member load is measured from the "start" end of the member in the direction of the positive X axis. The load location may be specified either as absolute distance from the start or as nondimensional fractions of the total length. Distributed member loads are specified as force per unit length along the local X axis.

The format of the MEMBER LOADS command is shown:

```
MEMBER LOADS (direction specs) (type specs)
Member number (direction specs) (type specs)
                    * ⎡ FORCE    X ⎤
                      ⎢          Y ⎥
                      ⎢          Z ⎥
direction specs =     ⎨            ⎬ (Global (Projected))
                      ⎢ MOMENT   X ⎥
                      ⎢          Y ⎥
                      ⎣          Z ⎦

              * ⎡ CONCENTRATED (FRACTIONAL) [P ]V₁ [L ]V₂                    ⎤
type specs =    ⎨ UNIFORM      (FRACTIONAL) [W ]V₃ ( [LA ]V₄ [LB ]V₅ )       ⎬
                ⎣ LINEAR       (FRACTIONAL) [WA]V₆ [WB]V₇ ( [LA ]V₈ [LB ]V₉ )⎦
```

where V_1 = magnitude of concentrated load

V_2 = distance from start of member to concentrated load

V_3 = magnitude of uniform force per unit length

V_4, V_5 = distance from start of member to start and end, respectively, of uniform load

V_6, V_7 = magnitude of force per unit length at start and end, respectively, of continuous load

V_8, V_9 = distance from start of member to start and end, respectively, of continuous load

The following examples illustrate the use of the three types of loading commands possible in the MEMBER LOADS command:

```
UNITS INCH POUNDS
MEMBER LOADS
1 TO 5 FOR Y GLO CONC P −8000. L 5.
6 9 10 FOR Y UNI FR W −300. LA 0. LB 1.0
7 8 MOM Y LIN FR WA 1000. WB 600. LA 0.9 LB .5
```

Any number of member loadings may be applied to any given member, but each must be described with a separate statement. Unless GLOBAL has been specified, member loads are assumed to act along the axes of the local coordinate system.

Preliminary structural design is usually done by using engineering judgment to select member sizes, geometry, and boundary conditions. The STRUDL program is then requested to provide results in terms of member stresses, deflections, and support reactions. These results should be compared to the allowable values in order to evaluate the adequacy of the structure. If it is within the allowable values, the design is usually not changed (unless it is extremely overdesigned). However, if the allowable criteria are exceeded, then a modification must be made and a new analysis done to verify the new configuration.

Some versions of STRUDL can automatically check all members to a specified engineering code, such as the AISC or ASME NF-XVII codes, among others. The methods of requesting this option vary from program to program; therefore the user is referred to the user manual for the particular version.

The time spent verifying the adequacy of the members may be avoided, since many programs, including STRUDL, have an option to have the program select the member sizing based on the input data. The authors believe that this option is rarely cost-effective, because a good engineer can usually select correct member sizes within a few tries, while the computer program may go through many iterations before experiencing equal success.

The selection of the locations in a member at which stresses should be reported is of major importance in structural analysis, because stresses usually vary along the member length. Therefore the engineer should specify locations along the member length at points of anticipated maximum stress. Maximum stress locations usually occur at points of maximum moment in the beam element. It is usually best to request stresses to be checked at the start, end, and middle of each member as well as at any load point.

The locations on the members where the stresses are to be checked and reported are specified below. Usually all members are chosen to have stresses reported unless the structure is symmetrical. The general form of the command is as follows:

$$\underline{\text{LIST}}, \text{ output type}, \begin{Bmatrix} \text{ALL (Members)} \\ \text{MEMBER list} \end{Bmatrix} (\underline{\text{SECTION}} \text{ specs})$$

$$\text{output type} = * \begin{Bmatrix} \underline{\text{SECTION FORCE}} \\ \underline{\text{SECTION STRESS}} \\ \underline{\text{FORCE ENVELOPE}} \\ \underline{\text{MAXIMUM STRESS}} \\ \underline{\text{STRESS ENVELOPE}} \end{Bmatrix}$$

$$\underline{\text{Section specs}} = (\text{FRACTIONAL}) \begin{Bmatrix} \underline{\text{NS}} \text{ i } v_1 \cdots v_i \\ \underline{\text{DS}} \text{ } v_1, v_2 \text{ } (\underline{\text{NS}} \text{ i}) \end{Bmatrix}$$

where i = number of sections to consider

$v_1, \cdots v_i$
 = distance from member start for each section

v_1 (with
$\underline{\text{DS}}$) = distance to first section

v_2 (with
$\underline{\text{DS}}$) = increment to each succeeding section (if $\underline{\text{NS}}$ i is not specified, sections continue to the member end)

TABLE 9.4 Sample STRUDL Output (Note: All annotated lines are user-entered, others are program-generated)

STRUDL 'Test'	Give name to problem
TYPE PLANE FRAME	Define type of analysis
UNITS FEET	Specify units
JOINT COORD	Specify coordinates of joints along X and Y axes. Mark joints 1 and 4 as support points.
1 0. 0.	
2 0. 10. S	
3 0. 25.	
4 12. 0. S	
5 12. 10.	
6 12. 25.	
MEMBER INCIDENCES	Define member incidences, listing first the member number, then the start and end joints, respectively.
1 1 2	
2 2 3	
3 4 5	
4 5 6	
5 2 5	
6 3 6	
UNITS KIP INCH	
CONSTANT E 29000. ALL	Enter modulus of elasticity in kips per square inch.
MEMBER PROP	Enter member properties from the table "tubes." The tube sizes are specified by depth (in inches), width (in inches), and thickness (in sixteenths of an inch).
1 TO 4 TABLE 'TUBES' 'T12X6X8'	
5 6 TABLE 'TUBES' 'T8X8X6'	
LOAD 1	Describe loading:
JOINT LOADS	"Load 1" has two joint loads.
2 3 FORCE X 4.	
STIFFNESS ANALYSIS	Instruct STRUDL to perform analysis
OUTPUT DECIMAL 2	Specify output format;
OUTPUT BY MEMBER	Two decimal places; list output in member order;
LIST SECTION STRESS ALL SECTION FRACTIONAL NS 3 0. .5 1.	Calculate stresses at three locations on each member (start, middle, and end).

```
****************************
* RESULTS OF LATEST ANALYSES *
****************************

PROBLEM-TEST

ACTIVE UNITS INCH KIPS RAD FAHR SEC LBM

INTERNAL MEMBER RESULTS

MEMBER NORMAL STRESS

   MEMBER  1
      LOADING  1

DISTANCE    /--------------------------------------- STRESS ---------------------------------------/
FROM START   AXIAL    Y SHEAR    Z SHEAR    Y BENDING    Z BENDING    MAX NORMAL    MIN NORMAL
 0.0    FR    0.40     -0.33       0.0         0.0         -8.47         8.87         -8.07
 0.500        0.40     -0.33       0.0         0.0         -3.14         3.54         -2.74
 1.000        0.40     -0.33       0.0         0.0          2.18         2.58         -1.78

   MEMBER  2
      LOADING  1

DISTANCE    /--------------------------------------- STRESS ---------------------------------------/
FROM START   AXIAL    Y SHEAR    Z SHEAR    Y BENDING    Z BENDING    MAX NORMAL    MIN NORMAL
 0.0    FR    0.17     -0.17       0.0         0.0         -3.52         3.70         -3.35
 0.500        0.17     -0.17       0.0         0.0          0.45         0.63         -0.28
 1.000        0.17     -0.17       0.0         0.0          4.43         4.60         -4.25

   MEMBER  3
      LOADING  1

DISTANCE    /--------------------------------------- STRESS ---------------------------------------/
FROM START   AXIAL    Y SHEAR    Z SHEAR    Y BENDING    Z BENDING    MAX NORMAL    MIN NORMAL
 0.0    FR   -0.40     -0.33       0.0         0.0         -8.43         8.03         -8.83
 0.500       -0.40     -0.33       0.0         0.0         -3.13         2.73         -3.53
 1.000       -0.40     -0.33       0.0         0.0          2.16         1.76         -2.56

   MEMBER  4
      LOADING  1
```

TABLE 9.4 Sample STRUDL Output (Note: All annotated lines are user-entered, others are program-generated) *(Continued)*

```
  DISTANCE    /------------------------------------ STRESS -------------------------------/
FROM START   AXIAL    Y SHEAR   Z SHEAR   Y BENDING   Z BENDING   MAX NORMAL   MIN NORMAL

    0.0  FR  -0.17    -0.17      0.0        0.0        -3.54         3.37        -3.72
    0.500    -0.17    -0.17      0.0        0.0         0.44         0.27        -0.62
    1.000    -0.17    -0.17      0.0        0.0         4.43         4.26        -4.61

  MEMBER   5
  LOADING  1

  DISTANCE    /------------------------------------ STRESS -------------------------------/
FROM START   AXIAL    Y SHEAR   Z SHEAR   Y BENDING   Z BENDING   MAX NORMAL   MIN NORMAL

    0.0  FR  -0.18     0.60      0.0        0.0        10.16         9.97       -10.34
    0.500    -0.18     0.60      0.0        0.0         0.01        -0.18        -0.19
    1.000    -0.18    -0.60      0.0        0.0       -10.15         9.96       -10.33

  MEMBER   6
  LOADING  1

  DISTANCE    /------------------------------------ STRESS -------------------------------/
FROM START   AXIAL    Y SHEAR   Z SHEAR   Y BENDING   Z BENDING   MAX NORMAL   MIN NORMAL

    0.0  FR  -0.19     0.46      0.0        0.0         7.88         7.70        -8.07
    0.500    -0.19     0.46      0.0        0.0        -0.00        -0.18        -0.19
    1.000    -0.19     0.17      0.0        0.0        -7.89         7.70        -8.07

LIST FORCE REACTIONS                            Request internal member loads and support reactions
```

```
*************************
*RESULTS OF LATEST ANALYSES*
*************************

PROBLEM - TEST

ACTIVE UNITS  INCH KIPS RAD  FAHR SEC LBM

MEMBER FORCES
```

MEMBER	LOADING	JOINT	/------FORCES------			/------MOMENTS------/		
			AXIAL	SHEAR Y	SHEAR Z	TORSION	MOMENT Y	MOMENT Z
1	1	1	-6.36	4.01				382.92
		2	6.36	-4.01				98.70
2	1	2	-2.78	2.00				159.30
		3	2.78	-2.00				200.20
3	1	4	6.36	3.99				380.87
		5	-6.36	-3.99				97.52
4	1	5	2.78	2.00				160.20
		6	-2.78	-2.00				200.30
5	1	2	1.98	-3.58				-257.99
		5	-1.98	3.58				-257.72
6	1	3	2.00	-2.78				-200.20
		6	-2.00	2.78				-200.30

JOINT LOADS - SUPPORTS

JOINT	LOADING	/------FORCES------			/------MOMENTS------/		
		X FORCE	Y FORCE	Z FORCE	X MOMENT	Y MOMENT	Z MOMENT
1	GLOBAL 1	-4.01	-6.36				382.92
4	GLOBAL 1	-3.99	6.36				380.87

LIST DISPLACEMENTS ALL   Request displacements under load for all joints

TABLE 9.4 Sample STRUDL Output (Note: All annotated lines are user-entered, others are program-generated)(*Continued*)

```
********************************
*RESULTS OF LATEST ANALYSES*
********************************

PROBLEM - TEST
ACTIVE UNITS  INCH  KIPS  RAD  FAHR  SEC  LBM
JOINT DISPLACEMENTS - SUPPORTS
```

JOINT	LOADING	/------DISPLACEMENTS------//			------ROTATIONS------/		
		X DISP	Y DISP	Z DISP	X ROT	Y ROT	Z ROT
1	GLOBAL 1	0.0	0.0				0.0
4	GLOBAL 1	0.0	0.0				0.0

```
JOINT DISPLACEMENTS - FREE JOINTS
```

JOINT	LOADING	/------DISPLACEMENTS------//			------ROTATIONS------/		
		X DISP	Y DISP	Z DISP	X ROT	Y ROT	Z ROT
2	GLOBAL 1	0.21	0.00				-0.00
3	GLOBAL 1	0.68	0.00				0.00
5	GLOBAL 1	0.21	-0.00				-0.00
6	GLOBAL 1	0.68	-0.00				-0.00

```
FINISH
```
End problem.

SOURCE: Courtesy of McDonnell Douglas Information Systems.

The command for requesting the maximum member stresses (plus or minus) for any load case, at the start, middle, or end of a member, is as follows:

```
LIST MAX STR ALL SEC FRA NS 3 0. .5 1.
```

Other desired results may be requested by the following command:

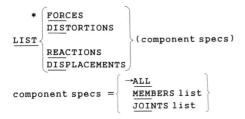

Forces are the internal member loads at the start and end of each member. Distortions are the relative displacements and rotations between the start and end joints of each member. Reactions are the support reactions. Displacements are the joint displacements of the structure under load. The results will be printed out for each individual loading case.

9.3.3 Sample structural analysis problem

To illustrate the use of the STRUDL program, a sample problem is provided in Table 9.4. The model of the frame, along with the desired loading cases, is shown in Fig. 9.19, in which members ① to ④ are TS 12 in × 6 in × ½ in thick, and members ⑤ and ⑥ are TS 8 in × 8 in × ⅜ in thick. For load case 1, 4 kips in the x direction is applied at joints 2 and

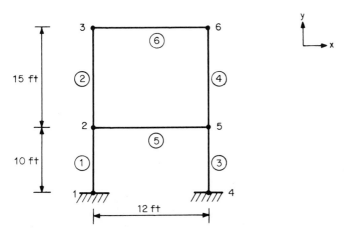

Figure 9.19 STRUDL model.

TABLE 9.5 MPDS Sample Pipe Stress Problem

Sample Input Screens

```
           DEFINING MATERIAL PROPERTY DATA

GROUP                                    1
MATERIAL NAME (8 CHARS. MAX.):           ss1                    ← Group numbers assigned in MPDS-MODEL
                                                                ← If the material name is in the data base, then data base properties are
                                                                    defaults
TEMPERATURES. . .(T1,T2,T3):            -100 70 800             ← Temperature table to define temperature-dependent material
                                                                    properties
ELASTIC MODULI . .(T1,T2,T3):           29e6 27.9e6 23.4e6      ← All entries are free-field, and may be scientific notation, decimal, or
EXPAN. COEF. . . .(T1,T2,T3):           5.65e-7 6.07e-7 7.65e-7     integer
ALLOW. STRESS . . (T1,T2,T3):           20000 19800 19400
POISSON RATIO. . . . . . . . :          .3                      ← Defaults to 0.3 if not entered
DENSITY. . . . . . . . . . . :          .283
OPERATING TEMPERATURE. . . :            800
OPERATING PRESSURE . . . . :            1000
PEAK PRESSURE. . . . . . . :            1200
REFERENCE TEMPERATURE. . . :            70

⟨ESC⟩–EXIT(NO CHANGE), ⟨F1⟩–CONTINUE, ⟨F10⟩–UPDATE ONLY         ← Status line defines key usage


                 SPECTRAL ACCEL. – LOAD CASE D1
GEOMETRY / DATE / TIME : SAMPLE / 10-DEC-1985 / 11:32:40

SPECTRUM NUMBER. . . :   1
DIRECTION COSINES. . :   1 0 0                                  ← Up to three spectra may be input
                                                                ← Specifies this spectrum to act in 1 direction only
FREQUENCY (1...5)  :   .1 3 5 8 10                              ← Free-field entry of spectra data pairs. All screen entries may be edited
AMPLITUDES (1...5) :   4 10 30 250 200                              using full-screen cursor movement before accepting the data into the
                                                                    program
FREQUENCY ( ...10) :   30 200
AMPLITUDES (6...10) :  85 10

FREQUENCY (11...15) :
AMPLITUDES (11...15) :

FREQUENCY (16...20) :
AMPLITUDES (16...20) :

⟨ESC⟩–EXIT(NO CHANGE), ⟨F1⟩–CONTINUE, ⟨F10⟩–UPDATE ONLY         ← Common commands used for all screens
```

Input Summary

NODAL DATA

NODE NUM.	KEYP NUM.	X-COORD (IN)	Y-COORD (IN)	Z-COORD (IN)	S.I.F.
1	1	0.0000E+00	0.0000E+00	0.0000E+00	
2	2	-0.4000E+02	0.0000E+00	0.0000E+00	1.40
3		-0.5550E+02	0.0000E+00	0.0000E+00	
4		-0.6000E+02	0.4500E+01	0.0000E+00	
5		-0.5550E+02	0.4500E+01	0.0000E+00	
6		-0.6000E+02	0.9000E+01	0.4500E+01	
7		-0.6000E+02	0.4500E+01	0.4500E+01	
8	14	-0.6000E+02	0.9000E+01	0.4000E+02	
9	6	-0.6000E+02	0.9000E+01	0.4600E+02	
10	7	-0.6000E+02	0.9000E+01	0.5200E+02	
11	8	-0.6000E+02	0.9000E+01	0.5800E+02	
12	9	-0.6000E+02	0.2100E+02	0.5200E+02	
13		-0.6000E+02	0.9000E+01	0.7540E+02	3.50
14		-0.6000E+02	0.1128E+02	0.7930E+02	1.95
15		-0.6000E+02	0.1350E+02	0.7540E+02	
16	12	-0.6000E+02	0.3498E+02	0.9300E+02	1.85
17	13	-0.6000E+02	0.6096E+02	0.1080E+03	
18	15	-0.6000E+02	0.1900E+02	0.4000E+02	

Absolute coordinates of all node points

SIF values entered to override those calculated by program

ELEMENT INPUT DATA

ELEM NUM.	ELEMENT TYPE	KEYP 1	KEYP 2	REF KEYP	NODE 1	NODE 2	NODE REF	MATE. GROUP	PROP. GROUP	FLEX FAC.
1	RUN	1	2		1	2	0	1	1	DEFLT
2	RUN	2			2	3	0	1	1	DEFLT
3	ELBOW				3	4	5	1	1	CODE
4	ELBOW				4	6	7	1	1	CODE
5	RUN		14		6	8	0	1	1	DEFLT
6	RUN	14	6		8	9	0	1	1	DEFLT
7	VALVE	6	7		9	10	0	2	2	DEFLT
8	VALVE	7	8		10	11	0	2	2	DEFLT
9	BEAM	7	9	6	10	12	9	3	3	DEFLT

Endpoints, materials, and cross-sectional properties must be defined for each element

TABLE 9.5 MPDS Sample Pipe Stress Problem (*Continued*)

Sample Input Screens

10	RUN	8		11	13	0	2	1	DEFLT
11	ELBOW			13	14	15	2	1	CODE
12	RUN	12		14	16	0	2	1	DEFLT
13	RUN	13		16	17	0	2	1	DEFLT
14	SPRING	15		8	18	0	1	6	DEFLT
15	MASS	9		12	0	0	1	5	DEFLT

C O N S T R A I N T   D A T A

NODE NUM.	KEYP. NUM.	GRAV. SUP.	THRM. SUP.	DYNM. SUP.	DIR	TRANSLATIONAL STIFFNESS (LBF/IN)	ROTATIONAL STIFFNESS (IN-LBF/RAD)
1	1	CCCCCC	CCCCCC	CCCCCC	X	RIGID	RIGID
					Y	RIGID	RIGID
					Z	RIGID	RIGID
2	2	FCFFFF	FCFFFF	FCEFFF	X	FREE	FREE
					Y	RIGID	FREE
					Z	0.600E+04	FREE
5		CCCCCC	CCCCCC	CCCCCC	X	RIGID	RIGID
					Y	RIGID	RIGID
					Z	RIGID	RIGID
7		CCCCCC	CCCCCC	CCCCCC	X	RIGID	RIGID
					Y	RIGID	RIGID
					Z	RIGID	RIGID
8	14	CFFFFF	CFFFFF	CFFFFF	X	RIGID	FREE
					Y	FREE	FREE
					Z	FREE	FREE
15		CCCCCC	CCCCCC	CCCCCC	X	RIGID	RIGID
					Y	RIGID	RIGID
					Z	RIGID	RIGID
16	12	FFFFFF	CFFFFF	FFFFFF	X	RIGID	FREE
					Y	FREE	FREE
					Z	FREE	FREE

Define location, function (active loading case), direction of action, and stiffness for each support.

Specify in each of 6 degrees of freedom either C (constrained), F (free), or E (elastic).

			CEEFFF	CEEFFF	CEEFFF	X		RIGID		FREE
17		13	CEEFFF	CEEFFF	CEEFFF	Y		0.500E+04		FREE
						Z		0.700E+04		FREE
18		15	CCCCCC	CCCCCC	CCCCCC	X		RIGID		RIGID
						Y		RIGID		RIGID
						Z		RIGID		RIGID

E L E M E N T   P R O P E R T Y   T A B L E S

GROUP	ELPAR(1)	ELPAR(2)	ELPAR(3)	ELPAR(4)	ELPAR(5)
1	0.3500E+01	0.2160E+00	0.0000E+00	0.0000E+00	0.0000E+00
2	0.3932E+01	0.4320E+00	0.9670E+01	0.0000E+00	0.0000E+00
3	0.6280E+03	0.3927E+05	0.3927E+05	0.0000E+00	0.0000E+00
4	0.1000E+00	0.0000E+00	0.0000E+00	0.0000E+00	0.0000E+00
5	0.2000E+03	0.0000E+00	0.0000E+00	0.0000E+00	0.0000E+00
6	0.8000E+04	0.0000E+00	0.0000E+00	0.0000E+00	0.0000E+00

L E G E N D - E L E M E N T   P R O P E R T Y   T A B L E S

TYPE	ELPAR(1)	ELPAR(2)	ELPAR(3)	ELPAR(4)	ELPAR(5)	ELPAR(6)
(BEAM)	AREA	IYY	IZZ	CKY	CKZ	
(CNST)	DEL_X					
(ELBO)	OD	WALL	IN_THICK	IN_DENS		
(MASS)	WEIGHT	RAD_GYRA				
(RUN)	OD	WALL	IN_THICK	IN_DENS		
(SPRG)	STIFF					
(VALV)	OD	WALL	WEIGHT	RAD_GYRA		

Cross-sectional properties referenced during element input.

TABLE 9.5 MPDS Sample Pipe Stress Problem (*Continued*)

Sample Problem Output

```
**** MICRO PIPING DESIGN SYSTEM-MPDS          ****
**** SAMPLE PROBLEM                           ****
**** ASME EQUATIONS 9 AND 11 CODE CHECKS      ****

LOAD CASE NO.  1  NAME: 61

STATIC ANALYSIS - DISPLACEMENTS
```

NOD	KEY	- UX - (IN)	- UY - (IN)	- UZ - (IN)	- ROT X - (RAD)	- ROT Y - (RAD)	- ROT Z - (RAD)
1	1	-.000	.000	.000	.00000	.00000	-.00000
2	2	-.000	.000	.010	.00031	.00039	.00009
3		-.000	-.003	.016	.00043	.00035	.00018
4		-.001	-.003	.020	.00049	.00025	.00021
5		.000	.000	.000	.00000	.00000	.00000
6		-.001	-.006	.022	.00056	.00010	.00024
7		.000	.000	.000	.00000	.00000	.00000
8	14	.000	-.025	.022	.00055	.00002	.00018
9	6	.000	-.029	.022	.00051	.00003	.00016
10	7	.000	-.032	.022	.00046	.00003	.00016
11	8	.001	-.034	.022	.00039	.00004	.00016
12	9	.001	-.038	.022	.00004	.00005	.00012
13		.001	-.037	.022	-.00024	.00008	.00010
14		.000	-.032	.012	-.00054	.00009	.00008
15		.000	.000	.000	.00000	.00000	.00000
16	12	-.000	-.022	-.004	-.00069	.00010	.00007
17	13	-.000	-.032	.028	-.00046	.00003	.00016
18	15	.000	.000	.000	.00000	.00000	.00000

← Global coordinate system

← Keypoint numbers are user-defined, while nodes are automatically defined by the program

```
****  MICRO PIPING DESIGN SYSTEM-MPDS        ****
****  SAMPLE PROBLEM                          ****
****  ASME EQUATIONS 9 AND 11 CODE CHECKS    ****

DESIGN CHECK RESULTS

DESIGN CHECK EQUATION : S-HEX         RANGE OF : T1
```

← Equation 11 (sustained + expansion). If more than one expansion case were specified, each possible range would be treated separately

← For SIF, greater of user input or program calculated for element (1.0 for pipe element) is used

← SIF automatically calculated by program per ASME code for elbow element

← Stresses N/A for valve, beam, and spring elements, but these elements are still included in the listing for continuity

EL. NO.	TYP	NODE	KEYP	SIF	INTERNAL PRESSURE (PSI)	SH TEMP (F)	SC TEMP (F)	ALLOW. STRESS (PSI)	CALC. STRESS (PSI)	STRESS RATIO
1	R	1	1	1.0	1000.0	800.	70.	49000.0	4979.5	0.10
		2	2	1.4	1000.0	800.	70.	49000.0	4211.9	0.09
2	R	2	2	1.4	1000.0	800.	70.	49000.0	4211.9	0.09
		3		1.0	1000.0	800.	70.	49000.0	4320.1	0.09
3	E	3		1.8	1000.0	800.	70.	49000.0	4941.2	0.10
		4		1.8	1000.0	800.	70.	49000.0	5122.0	0.10
4	E	4		1.8	1000.0	800.	70.	49000.0	5121.8	0.10
		6		1.8	1000.0	800.	70.	49000.0	4805.0	0.10
5	R	6		1.0	1000.0	800.	70.	49000.0	4225.8	0.09
		8	14	1.0	1000.0	800.	70.	49000.0	4365.6	0.09
6	R	8	14	1.0	1000.0	800.	70.	49000.0	4365.6	0.09
		9	6	1.0	1000.0	800.	70.	49000.0	4702.2	0.10
7	V	9	6	1.0	900.0	800.	70.	N/A	N/A	N/A
		10	7	1.0	900.0	800.	70.	N/A	N/A	N/A
8	V	10	7	1.0	900.0	800.	70.	N/A	N/A	N/A
		11	8	1.0	900.0	800.	70.	N/A	N/A	N/A
9	B	10	7	1.0	0.0	800.	70.	N/A	N/A	N/A
		12	9	1.0	0.0	800.	70.	N/A	N/A	N/A
10	R	11	8	1.0	900.0	800.	70.	52392.9	4652.2	0.09
		13		3.5	900.0	800.	70.	52392.9	6884.6	0.13
11	E	13		3.5	900.0	800.	70.	52392.9	6884.6	0.13
		14		2.0	900.0	800.	70.	52392.9	5037.4	0.10

TABLE 9.5 MPDS Sample Pipe Stress Problem (*Continued*)

EL. NO.	TYP	NODE	KEYP	SIF	INTERNAL PRESSURE (PSI)	SH TEMP (F)	SC TEMP (F)	ALLOW. STRESS (PSI)	CALC. STRESS (PSI)	STRESS RATIO
12	R	14		2.0	900.0	800.	70.	52392.9	5037.4	0.10
		16	12	1.9	900.0	800.	70.	52392.9	4470.6	0.09
13	R	16	12	1.9	900.0	800.	70.	52392.9	4470.6	0.09
		17	13	1.0	900.0	800.	70.	52392.9	2985.6	0.06
14	S	8	14	1.0	1000.0	800.	70.	N/A	N/A	N/A
		18	15	1.0	1000.0	800.	70.	N/A	N/A	N/A

```
**** MICRO PIPING DESIGN SYSTEM - MPDS        ****
**** SAMPLE PROBLEM                            ****
**** ASME EQATIONS 9 AND 11 CODE CHECKS        ****

LOAD CASE NO. 1 NAME: 61
STATIC ANALYSIS - SUPPORT REACTIONS
```

KEY PT.	NODE NO.	- FX - (LBF)	- FY - (LBF)	- FZ - (LBF)	- MX - (IN-LBF)	- MY - (IN-LBF)	- MZ - (IN-LBF)
1	1	-17.14	3.64	31.01	424.38	1307.11	-160.32
2	2	0.00	-81.04	0.00	0.00	0.00	0.00
14	8	19.09	0.00	0.00	0.00	0.00	0.00
13	17	-1.95	-110.24	-31.01	0.00	0.00	0.00
15	18	0.00	-202.35	0.00	0.00	0.00	0.00
TOT. LOAD		0.00	-389.99	0.00			

← For GRAV support scheme, thermal-only support at keypoint 12 and Z-snubber at keypoint 2 are not active

```
****  MICRO PIPING DESIGN SYSTEM-MPDS      ****
****  SAMPLE PROBLEM                       ****
****  ASME EQUATIONS 9 AND 11 CODE CHECKS  ****
```

DESIGN CHECK SUPPORT LOAD SUMMARY

NODE	KEYP	LOAD TYPE	RESULT	X-AXIS	Y-AXIS	Z-AXIS
1	1	SUST+OCCA+EXPN	+FORCE	140.	40.	77.
		SUST+EXPN	+(LB)	53.	4.	31.
		SUST	−	−17.	4.	31.
		SUST+EXPN	−	−17.	−8.	−15.
		SUST+OCCA+EXPN	−	−104.	−44.	−61.
1	1	SUST+OCCA+EXPN	+MOMENT	1563.	5570.	462.
		SUST+EXPN	+(LB-IN)	424.	1307.	0.
		SUST	−	424.	1307.	−160.
		SUST+EXPN	−	0.	−166.	−160.
		SUST+OCCA+EXPN	−	−769.	−4429.	−628.
2	2	SUST+OCCA+EXPN	+FORCE	0.	37.	249.
		SUST+EXPN	+(LB)	0.	0.	0.
		SUST	−	0.	−81.	0.
		SUST+EXPN	−	0.	−81.	0.
		SUST+OCCA+EXPN	−	0.	−187.	−249.
8	14	SUST+OCCA+EXPN	+FORCE	121.	0.	0.
		SUST+EXPN	+(LB)	19.	0.	0.
		SUST	−	19.	0.	0.
		SUST+EXPN	−	−87.	0.	0.
		SUST+OCCA+EXPN	−	−188.	0.	0.

← Program calculates worst load combinations, considering occasional load case as both + and −

All expansion load cases are automatically enveloped if more than one is specified

TABLE 9.5 MPDS Sample Pipe Stress Problem (*Continued*)

NODE	KEYP	LOAD TYPE	RESULT	X-AXIS	Y-AXIS	Z-AXIS
16	12	SUST+OCCA+HEXPN	+FORCE	58.	0.	0.
		SUST+HEXPN	+(LB)	58.	0.	0.
		SUST	—	0.	0.	0.
		SUST+HEXPN	—	0.	0.	0.
		SUST+OCCA+HEXPN	—	0.	0.	0.
17	13	SUST+OCCA+HEXPN	+FORCE	10.	0.	92.
		SUST+HEXPN	+(LB)	0.	0.	15.
		SUST	—	-2.	-110.	-31.
		SUST+HEXPN	—	-25.	-110.	-31.
		SUST+OCCA+HEXPN	—	-37.	-171.	-108.
18	15	SUST+OCCA+HEXPN	+FORCE	0.	0.	0.
		SUST+HEXPN	+(LB)	0.	0.	0.
		SUST	—	0.	-202.	0.
		SUST+HEXPN	—	0.	-244.	0.
		SUST+OCCA+HEXPN	—	0.	-436.	0.

SOURCE: Courtesy of Impell Corporation.

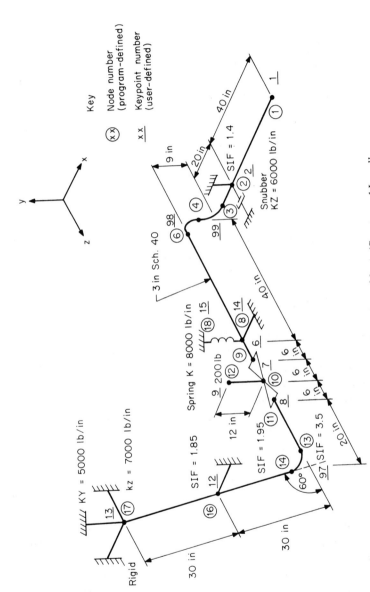

Figure 9.20 Micro piping design system (MPDS), sample problem. (*Courtesy of Impell Corporation.*)

3. The STRUDL input has been annotated to assist the engineer in the use of the most important STRUDL commands.

9.3.4 Sample pipe stress problem

To give the reader some familiarity with pipe stress analysis programs, several pages from a sample problem are provided in Table 9.5. The analysis is performed by using Micro Piping Design System (MPDS), a microcomputer-based program marketed by Impell Corporation.

Pipe stress programs share many features with structural analysis programs, in that the same types of data must be entered by the user and the same types of internal calculations are done. The differences are basically those of format, which varies slightly from program to program. One possible format difference may be that the system geometry may be specified by entering the incremental coordinates of each member (pipe segment) rather than the absolute coordinates of each node point. Also, the pipe segments may not be numbered (as structural members usually are), but rather identified by their start points and endpoints.

The sample problem input summary and output have been annotated to help the reader follow the program. This problem has been included only to familiarize the reader with these types of programs, since it is beyond the scope of this book to present a detailed account of pipe stress program input methods. The model for this problem is shown in Fig. 9.20.

The sample problem illustrates the use of a range of element types (straight pipe, elbow, 60° bend, valve, etc.), materials (three types of stainless steel), flexible and rigid restraints, both static and dynamic loading cases, and code compliance reporting.

The data is entered to this program through data screens, samples of which (those for entering material parameters and response spectra values) are shown in Table 9.5. The input data required is summarized according to type: node point coordinates and element start points and endpoints must be entered in order to define the system geometry; additional information entered here concerns the stress intensification factors (SIF) at each node point and the cross-sectional and material properties of each element. These properties are entered by referencing defined shape and material tables. Next, constraints on the system, in terms of pipe supports, must be entered. The information required here would be the function, the direction of action, and the stiffness of each support.

Output of typical results is also shown. This program, like most others, will calculate and display the displacements and rotations of each node point, the stress in each pipe segment (as well as the ratio of the stress to the maximum stress allowable by code), and the load on each support, all during each requested load case. To determine the maximum pipe support design load, a support load summary automatically combines the loads from each of the individual load cases.

Index

Access dimensions for piping layout, 47–48
Aesthetics in steel sizing, 205, 207
Air preheater, 9, 10
American Institute of Steel Construction (AISC), 19, 21, 22, 27–29, 211, 221, 234, 240
American Men and Women of Science, 11
American National Standards Institute (ANSI), 6, 19, 22–23, 36, 38, 61, 62, 69, 70, 72, 73, 75–77, 85, 88, 104, 109, 116, 292
  (*See also* Codes, *B31 series*)
American Nuclear Society (ANS), 37
American Society of Heating, Refrigeration, and Air Conditioning Engineers (ASHRAE), 30
American Society of Mechanical Engineers (ASME), 6, 18–20, 22–30, 38, 58, 61, 62, 68, 69, 73, 76, 77, 82, 85, 88, 120–121, 193, 205, 206, 211, 240, 292
  (*See also* Boiler and Pressure Vessel Code)
American Society for Testing and Materials (ASTM), 29, 213
American Standards Association (ASA), 6, 19, 22
  (*See also* Codes, *B31 series*)
American Welding Society (AWS), 19, 22, 31, 246–247
  welding symbols, table, 246–247
Amplification factor, 120
Analog computers, 274–275
Anchor bolts, 187–189, 194, 205
  evaluation of, 241, 245, 248–253
  reduction factor for, 245, 248–249
  (*See also* Baseplates)
Anchor movement, stresses due to, 84

Anchors, 57–58, 153, 181–182
Applicable documents, 17–37
As-built configuration, 210
As-built drawings, 42
Atomic Energy Commission (AEC), 31, 35
Atomic Industrial Forum (AIF), 37
Axial stress, 212–215, 218
Axial support, 57, 160, 179, 181

Bandwidth, 295, 297
Baseplates, 176, 177, 187–189, 204, 205, 211
  evaluation of, 241, 244–245, 248–253, 265–267
  (*See also* Anchor bolts; Embedded plates)
Basic Building Code (BBC), 31
Beam formulas, table, 226–227
Bearing lugs, 180–181
Bellows, 99–101
Bending diagram, 218–219, 222–223
Bending moment, 218
Bending stress, 218–220
Beta angle (STRUDL parameter), 304–305
Bill of materials, 208–209
Boiler, 5, 8–9, 42
Boiler and Pressure Vessel Code, ASME, 18, 22, 24–28, 30, 38, 58, 61, 62, 68, 76, 78–84, 205, 206, 240
  Subsection NB (Class 1), 24, 27, 29, 58, 62, 63, 78–82, 120
  Subsection NC (Class 2), 24, 27, 29, 58, 62, 63, 82–84, 120
  Subsection ND (Class 3), 24, 27, 29, 62, 82–84, 120
  Subsection NF (component supports), 24, 27, 193, 206

327

328 Index

Boiler explosions, 18–19
Boiler feed pump, 9, 10, 13
Boiling-water reactor (BWR), 12–13, 16
British Standards Institute, 36
B31 series (*see under* Codes)
Building codes, 31, 114–115
Building Officials and Code Administrators International (BOCA), 31

Calculations (*see* Computer calculations; Manual calculations)
Central processing unit (CPU), 280–281
Checking of design, 208–209
Checklist for design, 193
Chemical Plant and Petroleum Refinery Piping Code, B31.3, 72, 75–76, 292
Chemical process plant, 1
Circulating pump, 10, 14
Clamps, 155, 159–160, 174–176, 225, 256
Classes of nuclear piping:
    Class 1 (*see* Boiler and Pressure Vessel Code, Subsection NB)
    Class 2 (*see* Boiler and Pressure Vessel Code, Subsection NC)
    Class 3 (*see* Boiler and Pressure Vessel Code, Subsection ND)
Clevis hangers, 155, 159, 225
Code of Federal Regulations (CFR), 26–27, 31–32, 38, 115, 117
Code case, 20
Code case N-411, 120–121
Code stamps, ASME, table, 26
Codes, 17–31, 85, 86
    applicable edition of, 20–21
    Boiler and Pressure Vessel Code (*see* Boiler and Pressure Vessel Code)
    B31 Pressure Piping Code, 6, 22–23, 62, 69, 86
    B31.1 Power Piping Code, 70–72, 109–112, 292
    B31.3 Chemical Plant and Petroleum Refinery Piping Code, 72, 75–76, 292
    B31.7 Nuclear Power Piping Code, 76–77
    B31.8 Gas Transmission and Distribution Piping Code, 77–78, 86, 292
    history of, 18–19
    jurisdictional boundaries between, 27–29
    theoretical basis for, 62–68
Coefficient of friction, 160

Coefficient of thermal expansion, 128–129
Cold load of variable spring, 162–164
Cold springing, 66
Column design, 214–215, 218
Combining loads, 124–125, 150, 202–203
Components of thermoelectric plants, 8–10
Composite drawing, 42, 52–53, 195, 197, 199
Compression, 214–215, 218
Computer(s), 273–326
    analog, 274–275
    central processing unit (CPU), 280–281
    characteristics of, 274–277
    digital, 275, 277
    input/output devices, 277–280
    mainframe, 274–276, 285
    microcomputer, 274–276
    minicomputer, 274–276
    random-access memory (RAM), 277–278, 281, 283
    raster screen, 282–283
    stroke (vector) writing, 282–283
    super, 274
Computer-aided design and drafting (CADD), 273, 281–285
Computer-aided engineering (CAE), 273, 285–291
    (*See also* Computer calculations)
Computer calculations, 150, 202
    structural analysis using, 285–326
        element types, 288–289
        flexibility method, 290
        stiffness method, 290–291
    (*See also* Pipe stress analysis; structural analysis; STRUDL)
Conceptual design of pipe support, 203–205
Condensate pump, 8, 13, 14
Condenser, 8, 9, 13
Constant-spring supports, 164–165, 168, 170–173
Coordinate systems, 296–297
Corrosion, 44, 46, 152
Critical damping, 120
Cross-sectional properties, table, 216–217
Cumulative damage theory, 67–68

Damping, 118–122
    critical, 120
    for OBE and SSE, 120–121

Deflection, calculation of, 222–224, 267
Design checklist, 193
Design documents, 41–45, 193–195
Design drawing, 49–52, 208–209
Design guide, 20, 42
Design loads, 85–150, 202–203
    expansion loads, 128–149
    load combination, 150
    occasional loads, 102–128
    sustained loads, 86–102
Design specification, 41, 42, 44
Design temperature, 86, 151, 159
Detail design, 205–208
Detail drawing, 208–209
Determinate analysis, 234–236
Determination of thermal movements, 134–135, 147
Deutsches Institut für Normung (DIN), 36
Digital computer, 275, 277
Direction of support, 57, 153, 199, 201
    (See also Axial support; Lateral support; Vertical support)
Distance to first rigid support, calculating, 147–149, 201
Drag coefficient, 105–106
Dynamic analysis (see Modal analysis; Time history analysis)
Dynamic load factor (DLF), 110–112
Dynamic loading (see Occasional loads)

Earthquakes:
    map of earthquake zones, 116
    modified Mercalli scale, 113–114
    operating basis earthquake (OBE), 68, 117, 120, 121
    safe shutdown earthquake (SSE), 69, 117, 120, 121
    (See also Seismic loads; Seismic report)
Effective length of column, 214–215
Elevation view, 49–50
Embedded plates, 187–188, 199, 204
    evaluation of, 241, 244–245, 258
Emergency service level, 68, 69, 160
End conditions of beam, 234–235
Endurance limit, 67
Energy Reorganization Act, 31
Energy Research and Development Administration (ERDA), 31
Expansion of piping, 46, 128, 130–131
Expansion joints:
    bellows, 99–101

Expansion joints (*Cont.*):
    slip joint, 99, 101–102
Expansion loads, 71, 75, 78, 84, 86, 128–149, 202–203
    accommodation of thermal movements, 46, 156–158, 201
        using snubbers, 182–185
        using variable springs, 161–169
    determination of thermal movements, 134–135, 147
    expansion loops, 134–135
    friction caused by thermal movements, 160–161
    location of supports, 147–149
    stress calculation: charts, 136–145
        guided-cantilever method, 129, 132–134

Failure theories, 62–63
Fans, 9
Fatigue, 66–68
Faulted service level, 68, 69, 150
Feedwater heaters, 8, 9, 13
Finite element analysis, 287–290
Fixed connection, 234–235
Flexibility method of computerized structural analysis, 290
Flexible baseplate analysis, 211, 245, 250–252
Flexible range of response spectrum, 121, 122
Flexure formula, 219–220
Flow diagram, 41–43
Footprint loads, 208
Fossil fuel, 1, 3, 10–16
Frames, 178–179
    for multiple pipe supports, 46, 189–190
Free thermal analysis, 210
Friction, 160–161, 268–272
Function of support, 57–58, 199, 201, 203

Gas Transmission and Distribution Piping Code, B31.8, 77–78, 86, 292
General Agreement on Tariffs and Trade (GATT), 17
Generator, 8, 11, 13, 14
Guide (document), 20, 42
Guide (pipe support), 134–135
Guided-cantilever method, 129, 132–134
Gust factor, 106

330 Index

Hangers:
  clevis, 155, 159, 225
  rod, 57, 155–158
Hardware (*see* Pipe support hardware)
History:
  of codes, 18–19, 23–24
  of piping engineering, 5–7
Hooke's law, 212–213
Hot load of variable spring, 162–164
Hotwell, 8, 9

In-service inspection, 26, 42, 47, 58–60, 150, 204–205
Indeterminate analysis, 236–238
Industry standards, 17–20, 194
Input/output devices of computer, 277–280
Inspection, 152
  (*See also* In-service inspection)
Inspection and enforcement directives, NRC, 32–34
  IE Bulletin 79-14, 34
Institute of Nuclear Power Operations (INPO), 37
Insulation, 151–153, 159–161, 195
International Atomic Energy Agency (IAEA), 36–37
International Conference of Building Officials, 31
International Electrotechnical Commission (IEC), 36
International Standardization Organization (ISO), 36
International standards, 36
Isometric drawing, 41, 42, 53–58, 194, 199–201

Jurisdictional boundaries between codes, 27–29

Lateral support, 57, 179, 197, 198, 201
Layout (*see* Piping layout)
Licensee event reports (LERs), 35
Light-water reactor (LWR), 12
Limit stop, 134, 135
Line contact, 207, 228–229
Line list, 41, 43–45, 195
Line number, 43–44
Load capacity data sheet, 159–160, 194, 232–233, 256–259

Load combination, 124–125, 150, 202–203
Local stress, evaluation of, 207, 211, 225, 228–232
  line contact, 228–229, 268–270
  lugs, 229–231, 269–270
  saddles, 225, 228
  trunnions, 231–232, 261
Locations of supports, selection of, 55, 88–91, 103, 147–149, 195–198, 201
Loss-of-coolant accident (LOCA), 102
Lugs, 179–181, 197, 198, 229–231, 268–270
  bearing, 180–181
  shear, 181

Manual calculations, 211–272
  sample problems, 253–272
  for strength of materials, 212–224
  for support analysis, 225–253
  (*See also* Local stress; Structural analysis)
Manufacturers Standardization Society (MSS) of the valve and fitting industry, 23, 88–89
Maps:
  of earthquake zones, 116
  of wind velocities, 104
Massachusetts Institute of Technology, 285, 293
Maximum principal stress theory, 62–63
Maximum shear stress theory, 62–63
Micro Piping Design System (MPDS), sample output, 316–326
Modal analysis, 117–125
Model of plant, 42, 52–53, 198, 199
Modified Mercalli scale, 113–114
Modulus of elasticity, 213
Moment distribution, 238–239, 265
Multiple pipe supports, 46, 189–190

National Board of Boiler and Pressure Vessel Inspectors, 26, 30
National Fire Protection Association (NFPA), 37
Natural frequency, 120, 121, 123–124, 126, 127
  of piping system, determining, 123–124, 126, 127
Natural gas, 11
Naval Research Laboratory, 125

Index   331

NB, Subsection (*see* Boiler and Pressure Vessel Code, Subsection NB)
NC, Subsection (*see* Boiler and Pressure Vessel Code, Subsection NC)
ND, Subsection (*see* Boiler and Pressure Vessel Code, Subsection ND)
NF, Subsection (*see* Boiler and Pressure Vessel Code, Subsection NF)
Normal service level, 68, 150
North arrow on piping drawings, 51–52
Nuclear design, 115, 117
Nuclear piping classes, 24
  (*See also* Boiler and Pressure Vessel Code)
Nuclear power, 6, 10–12, 22, 24, 26, 31–37, 76–77, 112, 115–117, 206, 209
  (*See also* U.S. Nuclear Regulatory Commission)
Nuclear Power Piping Code, B31.7, 76–77
  (*See also* Boiler and Pressure Vessel Code)
Nuclear reactor:
  boiling-water reactor (BWR), 12–13, 16
  light-water reactor (LWR), 12
  pressurized-water reactor (PWR), 12, 14–16
Nuclear Regulatory Commission (*see* U.S. Nuclear Regulatory Commission)
NUREG reports, 34–35

Occasional loads, 68–69, 71, 75, 83, 86, 102–128
  (*See also* Relief valve discharge loads; Seismic loads; Vibration; Wind loads)
Occupational Safety and Health Administration (OSHA), 30–31
Oil, 10, 11
Operating basis earthquake (OBE), 68, 117, 120, 121
Organization chart of project, 40–41
Oscillators, 118–121

Pacific Area Standards Congress (PASC), 36
Peak spreading of response spectrum, 122–123
Peak stress, 63, 64, 66
Pinned connections, 234–235
Pipe Fabrication Institute, 30

Pipe rupture, 42, 102
Pipe straps, 178, 225
Pipe stress analysis, 7, 61–85, 150, 198–202, 291–292, 316–326
  program capabilities, 291–292
  requirements for, 70–85
  sample output, 316–326
Pipe support hardware, 151–190, 203–204
  evaluation of, 211, 232
Piping and instrumentation drawing, 41, 42
Piping codes (*see* Codes)
Piping design documents, 41–45, 193–195
Piping drawing, 41, 49–52, 194
Piping engineer, duties of, 40–41
Piping layout, 46–48
  access dimensions, 47–48
Piping system, 1–2, 85–86
Plan view, 49–50
Power Piping Code, B31.1, 70–72, 109–112, 292
Power plants, 1, 3, 7–11
  components of, 8–10
Precipitator, 9, 10
Pressure loads, 98–102
Pressure Piping Code, B31, 6, 22–23, 62, 69, 86
Pressurized-water reactor, 12, 14–16, 29
Primary stress, 63, 77–79
Project, organization chart of, 40–41
Project schedule, 191–192
Properties:
  of cross sections, table, 216–217
  of weld groups, table, 243–244
Prying force in baseplate, 211, 250–252

Quality assurance, 38–39, 41

Radius of gyration, 214
Raster screen, 282–283
Reactor vessel, 13, 14, 29
Recommended practices, 18–20
Reduction factor for anchor bolts, 245, 248–249
Regulations, 3, 17–37
Regulatory guides, 32–33
  NRC Regulatory Guide 1.92, 125
Relief valve discharge loads, 109–113
Resonant range of response spectrum, 121, 122

Response spectrum, 118–127
  flexible range of, 121, 122
  peak spreading, 122–123
  resonant range of, 121, 122
  rigid range of, 121, 122, 125, 127
Reynolds number, 105–106
Right-hand rule, 296–297
Rigid baseplate analysis, 245, 250–251
Rigid range of response spectrum, 121, 122, 125, 127
Rigid supports, 57, 153–161, 169–182, 201, 206
  frames, 178–179, 189–190
  pipe straps, 178
  struts, 169, 174–176
  U-bolts, 176–178
  for weight loads, 153–161
Riser clamps, 155, 160
Rod hangers, 57, 155–158
Routing (see Piping layout)

Saddles, 154, 160–161, 225, 228
Safe shutdown earthquake (SSE), 69, 117, 120, 121
Safety analysis report (SAR), 35
Sample outputs:
  pipe stress, 316–326
  STRUDL, 310–315, 326
Scale model of plant, 42, 52–53, 198, 199
Schedule of project, 191–192
Screenhouse, 9, 10
Secondary stress, 63–66
Section modulus, 220
  of weld group, 240
Sectional properties, table, 216–217
Seismic loads, 112–127, 202–203
  combining, 124–125
  modal analysis, 117–125
  response spectrum (see Response spectrum)
  static analysis, 115, 125–127
  time history analysis, 117
  Uniform Building Code, 115, 125
  (See also Earthquakes)
Seismic report, 58
Self-springing, 66
Service levels, 68–69, 150
Shake down, 65–66
Shear diagram, 220, 222–223
Shear lugs, 181
Shear stress, 62–63, 212, 220–222
Shell-type anchor bolt, 188

Slenderness ratio, 214
Sliding support, 154, 160–161
Slip joint, 99, 101–102
Snubbers, 153, 169, 182–185, 197, 201, 262–263
  problems with, 182–185
Southern Building Code Congress International (SBCCI), 31
Spring (see Constant-spring supports; Variable-spring supports)
Spring constant:
  of support, 206–207
  of variable spring, 162–164
Standard Building Code (SBC), 31
Standard design, 6, 17–20, 194
Standard review plans (SRPs), 35, 36
Standard spans, 88–90
Standards, 17–20
  international, 36
Static analysis, 115, 125–127
Steam generator, 8, 10, 14, 16, 29
Steel design, 191, 207, 232, 234–240
  (See also Structural analysis)
Steel drawing, 194–196
Stiffness of support (see Spring constant)
Stiffness method of computerized structural analysis, 290–291
Strain, 65–66, 212–213
Strength of materials, 212–224
Stress(es):
  axial, 212–215, 218
  bending, 63, 65, 218–220
  compression, 214–215, 218
  limits for, 64–66
  normal, 212
  peak, 63, 64, 66, 81
  primary, 63, 77–79
  secondary, 63–66, 79
  shear, 62–63, 212, 220–222
  tension, 64–65, 213–214
  (See also Pipe stress analysis)
Stress analysis (see Pipe stress analysis)
Stress intensification factor, 71–76, 78, 84, 326
Stress intensity, 62
Stress isometric drawing, 42, 55–58
Stress range reduction factor, 71, 72, 75–76, 84
Stress report, 42, 58
Stress-strain curve, 65–66, 212–213
Stroke (vector) writing, 282–283
Structural analysis, 205–206, 234–239
  determinate analysis, 234–236

Structural analysis (*Cont.*):
  indeterminate analysis, 236–238
  moment distribution, 238–239
  (*See also* Computer calculations, structural analysis using; STRUDL)
Structural drawing, 194–196
STRUDL (structural design language), 285–286, 291–315, 326
  commands for, 298–309, 315
  sample output, 310–315, 326
Struts, 169, 174–176
Submarine piping, 2
Support conceptual design, 203–205
Support design process, 191–210
Support detail design, 205–208
Support hardware (*see* Pipe support hardware)
Support types, 57–58
  (*See also* Function of support)
Sustained loads, 68, 70–71, 75, 83, 86–102
  (*See also* Pressure loads; Weight loads)
Sway braces, 153, 186–187

Technical documents, 41–60, 193–195
Temperature considerations, 86, 151, 159
Temperature derating factor, 77
Tension, 213–214
Theoretical basis for piping codes, 62–68
Thermal expansion, 128, 130–131
Thermal loads (*see* Expansion loads)
Thermal movements (*see* Expansion loads)
Thermoelectric plants (*see* Power plants)
Three Mile Island, 11
Time history analysis, 117
Torsion, 220–222
Torsional resistivity, 221–222
Trunnions, 181, 231–232
Turbine, 8, 9, 13, 14, 42

U-bolts, 176–178
U.S. Nuclear Regulatory Commission, 30–38, 68, 120, 122
  inspection and enforcement directives, 32–34
    IE Bulletin 79-14, 34
  licensee event reports (LERs), 35
  NUREG reports, 34–35

U.S. Nuclear Regulatory Commission (*Cont.*):
  regulatory guides, 32–33
    Regulatory Guide 1.92, 125
  standard review plans (SRPs), 35, 36
Uniform Boiler and Pressure Vessel Society, 26, 29–30
Uniform Building Code, 31, 115, 125
Upset service level, 68–69, 150

Valve list, 44
Variability of variable spring, 162–164
Variable-spring supports, 161–169, 187, 201
  cold load, 162–164
  hot load, 162–164
  selection criteria for, 166–169, 253–254
  spring constant of, 162–164
  variability of, 162–164
Vector writing, 282–283
Vertical support, 57, 153
Vibration, 127–128, 186–187

Wedge-type anchor bolt, 188
Weight balancing, 90–98
Weight loads, 87–98, 202–203
  calculation of, 87–98
  standard spans, 88–90
  weight balancing, 90–98
Weight supports, 152–169
  rod hangers, 155–158
Weights of piping materials, table, 94–96
Welded attachments, 179–182, 207
  lugs, 179–181, 229–231
  saddles, 225, 228
  trunnions, 181, 231–232
Welding Research Council (WRC), 35–36, 229–232, 261
Welding symbols, table, 246–247
Welds, evaluation of, 240–244, 258, 261, 264–265, 272
  properties of, table, 243–244
Wind loads, 103–109
Wind velocity, map of, 104

Yield point (*see* Yield stress)
Yield stress, 63–66, 212–213

Zero-period acceleration (ZPA), 121, 127

ABOUT THE AUTHORS

PAUL RICHARD SMITH, a Registered Professional Engineer in Massachusetts and New York, has conducted through his own firm, PRS Energy Industries, Inc., more than 100 workshops on piping design for both professional and technically oriented lay audiences. He has designed piping and support systems for more than ten nuclear power plants in the United States. His experience also extends to the international arena. As a technical expert for Ebasco Services, he audited a large manufacturer of piping in Europe, and coordinated American and European piping designs for a nuclear power facility.

THOMAS J. VAN LAAN, a Registered Professional Engineer in New York and Connecticut, is president of Tetracom Services, a firm specializing in computer-aided engineering. He has worked for Ebasco Services, Inc., and Bechtel Power Corp. as project lead in the fields of pipe stress, pipe supports, and mechanical engineering. He also has worked for Auton/Intercomp, providing engineering support for the marketing of Dynaflex, a computerized pipe-stress program. In addition, he has conducted seminars in pipe stress for PRS Energy Industries, Inc.